BRITISH RAILWAYS POCK...

LOCC...

THIRTY-SECOND EDITION – 1994

The Complete Guide to all
BR-operated Locomotives

Peter Fox and
Richard Bolsover

PLATFORM
5

ISBN 1 872524 57 5

CONTENTS

FROM THE PUBLISHER

This book, formerly known as 'Motive Power Pocket Book', contains details of all BR-operated locomotives. This includes BR and SNCF locomotives which will work through the Channel Tunnel, but the 'Eurostar' trains owned by the state railway companies of Britain, France and Belgium and the stock owned by the Channel Tunnel Company is included in Railways Pocket Book No. 3 – DMUs and Channel Tunnel Stock. Information is updated to 3rd December 1993. We would like to thank all who have helped by contributing information or photographs.

Platform 5 Publishing are still the only company to provide the enthusiast or transport official with complete information on BR rolling stock. This book is, at the time of publication, the most up-to-date publication of its type on the market.

In the centre of this issue will be found details of our book club which features pre-publication discounts on most Platform 5 books (excluding our BR pocket books). It is hoped that enthusiasts will find this facility convenient and cost effective.

NOTES

The following notes are applicable to locomotives:

LOCOMOTIVE CLASS DETAILS

Principal details and dimensions are given for each class in metric units. Imperial equivalents are also given for power. Maximum speeds are still quoted in miles per hour since the operating department of BR still uses imperial units. Since the present maximum permissible speed of certain classes of locomotives is different from the design speed, these are now shown separately in class details. In some cases certain low speed limits are arbitrary and may occasionally be raised when necessary if a locomotive has to be pressed into passenger service.

Standard abbreviations used are:

ABB	ABB Transportation Ltd.	kN	kilonewtons
BR	British Railways	kV	kilovolts
BREL	British Rail Engineering Ltd.	kW	kilowatts
	(later BREL Ltd.)	lbf	pounds force
BRML	British Rail Maintenance Ltd.	mph	miles per hour
hp	horse power	RA	Route availability
km/h	kilometres/hour	t	tonnes

LOCOMOTIVE DETAIL DIFFERENCES

Detail differences which affect the areas and types of train which locos work are shown. Where detail differences occur within a class or part class of locos., these are shown against the individual loco number. Except where shown, diesel locomotives have no train heating equipment. Electric or electro-diesel locomotives are assumed to have train heating unless shown otherwise. Standard abbreviations used are:

a   Train air brakes only
c   Cab to shore radio-telephone fitted.
e   Fitted with electric heating apparatus (ETH).
r   Fitted with radio electronic token block equipment.
s   Slow speed control fitted (and operable).
t   Fitted with automatic vehicle identification transponders.
v   Train vacuum brakes only.
x   Dual train brakes (air & vacuum).
y   ETH equipped but equipment isolated.
+   Extended range locos with Additional fuel tank capacity compared with others in class.

NAMES AND ALLOCATIONS

All official names are shown as they appear on the locomotive i.e. all upper case or upper & lower case lettering.

(S) denotes stored serviceable and (U) stored unserviceable. Last known allocations of stored locomotives are shown, but readers should note that locomotives may not necessarily be stored at their home depots.

After the locomotive number are shown any notes regarding braking, heating etc., the livery code (in bold type), the pool code, the depot code and name if any. Locomotives which have been renumbered in recent years show the last number in parentheses after the current number. Where only a few locomotives in a class are named, these are shown in a separate table at the end of the class or sub-class.

Thus the layout is as follows:

No.	Old No.	Notes	Liv.	Pool	Depot	Name
47636	(47243)	+	**RX**	PXLB	CD	Restored

GENERAL INFORMATION ON BRITISH RAILWAYS' LOCOMOTIVES

CLASSIFICATION & NUMBERING

Initially BR diesel locomotives were allocated numbers in the 1xxxx series, with electrics allotted numbers in the 2xxxx series. Around 1957 diesel locomotives were allocated new four digit numbers with 'D' prefixes. Diesel electric shunters in the 13xxx series had the '1' replaced by a 'D', but diesel mechanical shunters were completely renumbered. Electric locomotives retained their previous numbers but with an 'E' prefix. When all standard gauge steam locomotives had been withdrawn, the prefix letter was removed.

In 1972, the present TOPS numbering system was introduced whereby the loco number consisted of a two-digit class number followed by a serial number. In some cases the last two digits of the former number were generally retained (classes 20, 37, 50), but in other classes this is not the case. In this book former TOPS numbers carried by converted locos. are shown in parentheses. Full renumbering information is to be found in the 'Diesel & Electric loco Register'. This is at present out of print, but a new edition may be published in the future.

Diesel locomotives are classified as "types" depending on their engine horsepower as follows:

Type	Engine hp.	Old Number Range		Current Classes
1	800 – 1000	D 8000 – D 8999		20
2	1001 – 1499	D 5000 – D 6499/D 7500 – D 7999	26, 31.
3	1500 – 1999	D 6500 – D 7499		33, 37.
4	2000 – 2999	D 1 – D 1999		47, 50.
5	3000 +	D 9000 – D 9499		56, 58, 59, 60.
Shunter	Under 300	D 2000 – D 2999		03.
Shunter	300 – 799	D 3000 – D 4999		08, 09.

Class 14 (650 hp diesel hydraulics) were numbered in the D95xx series.

Electric and electro-diesel locomotives are classified according to their supply system. Locomotives operating on a d.c. system are allocated classes 71 – 80 whilst a.c. or dual voltage locomotives start at Class starting at 81. Departmental locomotives which remain self propelled or which are likely to move around on a day to day basis are classified 97.

WHEEL ARRANGEMENT

For main line diesel and electric locomotives the system whereby the number of driven axles on a bogie or frame is denoted by a letter (A = 1, B = 2, C = 3 etc.) and the number of undriven axles is noted by a number is used. The letter 'o' after a letter indicates that each axle is individually powered and a + sign indicates that the bogies are intercoupled.

For shunters and steam locomotives the Whyte notation is used. The number of leading wheels are given, followed by the number of driving wheels and then the trailing wheels. Suffix 'T' on a steam locomotive indicates a tank locomotive, and 'PT' a pannier tank loco.

HAULING CAPABILITY OF DIESEL LOCOS

The hauling capability of a diesel locomotive depends basically upon three factors:

1. Its adhesive weight. The greater the weight on its driving wheels, the greater the adhesion and thus more tractive power can be applied before wheel slip occurs.

2. The characteristics of its transmission. In order to start a train the locomotive has to exert a pull at standstill. A direct drive diesel engine cannot do this, hence the need for transmission. This may be mechanical, hydraulic or electric. The current BR standard for locomotives is electric transmission. Here the diesel engine drives a generator or alternator and the current produced is fed to the traction motors. The force produced by each driven wheel depends on the current in its traction motor. In other words the larger the current, the harder it pulls.

As the locomotive speed increases, the current in the traction motors falls hence the *Maximum Tractive Effort* is the maximum force at its wheels that the locomotive can exert at a standstill. The electrical equipment cannot take such high currrents for long without overheating. Hence the *Continuous Tractive Effort* is quoted which represents the current which the equipment can take continuously.

3. The power of its engine. Not all of this power reaches the rail as electrical machines are approximately 90% efficient. As the electrical energy passes through two such machines (the generator/alternator and the traction motors), the *Power At Rail* is about 81% (90% of 90%) of the engine power, less a further amount used for auxiliary equipment such as radiator fans, traction motor cooling fans, air compressors, battery charging, cab heating, ETH, etc. The power of the locomotive is proportional to the tractive effort times the speed. Hence when on full power there is a speed corresponding to the continuous tractive effort.

HAULING CAPABILITY OF ELECTRIC LOCOS

Unlike a diesel locomotive, an electric locomotive does not develop its power on board and its performance is determined only by two factors, namely its weight and the characteristics of its electrical equipment. Whereas a diesel locomotive tends to be a constant power machine, the power of an electric locomotive varies considerably. Up to a certain speed it can produce virtually a constant tractive effort. Hence power rises with speed according to the formula given in section 3 above, until a maximum speed is reached at which tractive effort falls, such that the power also falls. Hence the power at the speed corresponding to the maximum tractive effort is lower than the maximum.

BRAKE FORCE

The brake force is a measure of the braking power of a locomotive. This is shown on the locomotive data panels so that railway staff can ensure that sufficient brake power is available on freight trains.

TRAIN HEATING EQUIPMENT

Electric train heating (ETH) is now the standard system in use on BR for loco-hauled trains. Locomotives which were equipped to provide steam heating have had this equipment removed or rendered inoperable (isolated). Electric heat is provided from the locomotive by means of a separate alternator on the loco., except in the case of and classes 33 and 50 which have a d.c. generator. The *ETH Index* is a measure of the electrical power available for train heating. All electrically heated coaches have an ETH index and the total of these in a train must not exceed the ETH power of a locomotive.

ROUTE AVAILABILITY

This is a measure of a railway vehicle's axle load. The higher the axle load of a vehicle, the higher the RA number on a scale 1 to 10. Each route on BR has an RA number and in theory no vehicle with a higher RA number may travel on that route without special clearance. Exceptions are made, however.

MULTIPLE AND PUSH-PULL WORKING

Multiple working between diesel locomotives on BR has usually been provided by means of an electro-pneumatic system, with special jumper cables connecting the locos. A coloured symbol is painted on the end of the locomotive to denote which system is in use. Class 47/7 used a time-division multiplex (t.d.m.) system which utilised the existing RCH (an abbreviation for the former railway clearing house, a pre-nationalisation standards organisation) jumper cables for push-pull working. These had in the past only been used for train lighting control, and more recently for public address (pa) and driver – guard communication. A new standard t.d.m. system is now fitted to all a.c. electric locomotives and other vehicles, enabling them to work in both push-pull and multiple working modes.

BR DIESEL LOCOMOTIVES

CLASS 03   BR SHUNTER   0-6-0

Built: 1960 at BR Doncaster Works.
Engine: Gardner 8L3 of 152 kW (204 hp) at 1200 rpm.
Transmission: Mechanical. Fluidrive type 23 hydraulic coupling to Wilson-Drewry CA5R7 gearbox with SCG type RF11 final drive.
Max. Tractive Effort: 68 kN (15300 lbf).

Brake Force: 13 t.	**Length over Buffers:** 7.92 m.
Weight: 31 t.	**Wheel Diameter:** 1092 mm.
Max. Speed: 28 mph.	**RA:** 1.

Formerly numbered 2079.

03079 v   NKJD   RY

CLASS 08   BR SHUNTER   0-6-0

Built: 1953-62 by BR at Crewe, Darlington, Derby, Doncaster or Horwich Works.
Engine: English Electric 6KT of 298 kW (400 hp) at 680 rpm.
Main Generator: English Electric 801.
Traction Motors: Two English Electric 506.
Max. Tractive Effort: 156 kN (35000 lbf).
Cont. Tractive Effort: 49 kN (11100 lbf) at 8.8 mph.

Power At Rail: 194 kW (260 hp).	**Length over Buffers:** 8.92 m.
Brake Force: 19 t.	**Wheel Diameter:** 1372 mm.
Design Speed: 20 mph.	**Weight:** 50 t.
Max. Speed: 15 or 20* mph.	**RA:** 5.

Non standard liveries:

08414 is 'D' with Rfd brandings and also carries its former number D 3529.
08500 is red lined out in black & white.
08519/730/867 are BR black.
08593 is Great Eastern blue lined out in red and also carries its former number D 3760.
08601 is London Midland & Scottish Railway black.
08629 is Royal purple.
08642 is London & South Western Railway black and also carries its former number D 3809.
08689 is 'D' with Railfreight general brandings.
08715 is in experimental dayglo orange livery.
08721 is blue with a red & yellow stripe ("Red Star" livery).
08883 is Caledonian blue.
08907 is London & North Western Railway black.
08933 is 'D' but with two orange cabside stripes.
08938 is grey and red.

n - Waterproofed for working at Oxley Carriage Depot.
z - Fitted with buckeye adaptor at nose end for HST depot shunting.

§ – Fitted with yellow flashing light and siren for working between Ipswich Yard and Cliff Quay.

Formerly numbered in series 3000 – 4192. 08600 was numbered 97800 whilst in departmental use between 1979 and 1989.

CLASS 08/0. Standard Design.

08388	a	F	FSNI	IM	08516	a	D	FSCK	KY
08389	a		FSNT	TI	08517	a		MSSX	MR (S)
08393	a	D	MSSS	SF	08519	a	O	MSSB	BY
08397	a	F	MSNA	AN	08523	x		MSSR	RG
08401	a	D	FSNI	IM	08525	x	F	FSCK	KY
08402	a	D	MSNA	AN	08526	x		MSSS	SF
08405	a	D	FSNI	IM	08527	x	D	MSSS	SF
08410	a	D	MSSA	BR	08528	x	D	MSSM	MR
08411	a		FSSM	ML	08529	x		MSSM	MR
08413	a	D	MSSR	RG	08530	x	D	MSSS	SF
08414	a * §	O	MSSS	SF	08531	x	D	MSSS	SF
08415	x		MSNA	AN	08534	x	D	MSNU	CL
08417	a	D	MSCH	SF	08535	x	D	MSNB	BS
08418	a	F	FSCD	DR	08536	x		MSNE	DY
08428	a		MSNB	BS	08538	x	D	MSSM	MR
08441	a		FSCN	TO	08540	x	D	MSSM	MR
08442	a	F	FSCD	DR	08541	x	D	MSSS	SF
08445	a		FSNX	IM	08542	x	F	MSSS	SF
08447	a		MSNX	CL (S)	08543	x	D	MSNB	BS
08448	a		MSNX	BS (S)	08561	x		FSSA	AY
08449	a		FSCN	TO	08562	x		MSCH	DR
08451	x		MSSW	WN	08567	x		MSSB	BY
08454	x		MSSW	WN	08568	x		FSSM	ML
08460	a	F	MSSO	OC	08569	x		MSNA	AN
08466	a	FO	FSNI	IM	08571	xz		FSSM	ML
08472	a		MSSX	SF	08573	x		MSSS	SF
08480	az		MSSO	OC	08575	x	BS	FSNL	NL
08481	x		FSWX	CF (U)	08576	x		MSSL	LA
08482	a	D	MSNA	AN	08577	x		FSNX	HT (S)
08483	a	D	MSSA	BR	08578	x	R	FSNH	HT
08484	a	D	MSSB	BY	08580	x		MSSM	MR
08485	a		MSNA	AN	08581	x		FSNT	TI
08489	a	F	MSNA	AN	08582	a	D	FSNY	TE (S)
08492	a		FSCN	TO	08585	x		MSNC	CD
08493	a		FSWK	CF	08586	a	F	FSSA	AY
08495	x		MSSM	MR	08587	x		FSNH	HT
08498	a		MSCH	SF	08588	xz	BS	FSNL	NL
08499	a	F	FSCK	KY	08593	x	O	MSSS	SF
08500	x	O	FSCD	DR	08594	x		MSSC	CA
08506	a		FSSM	ML	08597	x		FSCN	TO
08509	a	F	FSNT	TI	08599	x		MSNC	CD
08510	a		FSNT	TI	08600	a	D	NKJD	EH
08511	a		FSCN	TO	08601	x	O	MSNX	BS (S)
08512	a	F	FSCD	DR	08603	x		MSCH	BS
08514	a		FSCD	DR	08605	x		FSCK	KY

08607	x		FSCN	TO		08697	x		MSNE	DY
08610	x		MSNB	BS		08698	a		MSSX	SF (S)
08611	x		MSNX	LO (S)		08700	a		MSSX	SF (S)
08613	x		MSNX	AN (S)		08701	x	R	FSNH	HT
08615	x		MSNA	AN		08702	x		MSNC	CD
08616	x		MSNB	BS		08703	a		MSNX	AN (S)
08617	x		MSSW	WN		08705	a		MSCH	MR
08619	x		MSNX	LO (S)		08706	x		FSCK	KY
08622	x		FSSM	ML		08709	x		MSSS	SF
08623	x		FSCN	TO		08711	x		MSSC	CA
08624	x		MSNL	LO		08713	a		MSCH	MR
08625	x		MSSB	BY		08714	x		MSSC	CA
08627	a		MSSS	SF		08715	v	0	MSSS	SF
08628	x		MSSB	BY		08718	x		FSSM	ML
08629	x	0	MSSB	BY		08720	a	D	FSSM	ML
08630	x		FSSM	ML		08721	x	0	MSNL	LO
08632	x		FSNI	IM		08723	x		FSCN	TO
08633	x	RX	MSNC	CD		08724	x		MSSS	SF
08635	x		MSNC	CD		08730	x	0	FSSM	ML
08641	xz	D	MSSL	LA		08731	x		FSSM	ML
08642	x*	0	NKJD	EH		08734	x		MSNB	BS
08643	xz	D	MSSA	BR		08735	x		FSSM	ML
08644	xz	I	MSSL	LA		08737	x	F	MSNC	CD
08645	xz	D	MSSL	LA		08738	x	D	FSSM	ML
08646	x	F	FSWK	CF		08739	x		MSNC	CD
08648	x*	D	MSSL	LA		08740	x	F	MSSS	SF
08649	x	G	NKJD	EH		08742	x		MSNC	CD
08651	xz	D	MSSO	OC		08745	xz	BS	FSNL	NL
08653	x*		MSSO	OC		08746	x	D	MSNB	BS
08655	x*	F	MSSS	SF		08748	x §		MSSX	SF
08661	a		FSNX	NL (S)		08750	x		MSSS	SF
08662	x		FSCK	KY		08751	x		MSNB	BS
08663	a	D	MSSL	LA		08752	x	C	MSSS	SF
08664	x		FSWX	CF (S)		08754	x		FSSI	IS
08665	x		FSNI	IM		08756	x	D	FSWK	CF
08666	x		MSNX	LO (S)		08757	x	RX	MSSC	CA
08668	x		MSSA	BR		08758	x		MSSS	SF
08670	a		MSCH	SF		08762	x		FSSI	IS
08673	x	IO	MSNL	LO		08765	xn	D	MSNB	BS
08675	x	F	FSSA	AY		08767	x		MSSS	SF
08676	x		MSNL	LO		08768	x		MSNU	CL
08682	x		FSCD	DR		08770	a	D	FSWK	CF
08683	x		MSSB	BY		08772	x	G	MSSX	SF
08685	x		MSSC	CA		08773	x		FSCN	TO
08689	a	0	MSSS	SF		08775	x		MSSS	SF
08690	x		MSNE	DY		08776	x	D	FSCK	KY
08691	x	G	FSNT	TI		08780	x		FSWX	CF (S)
08692	x		MSNC	CD		08782	a		FSCK	KY
08693	x		FSSM	ML		08783	x		FSCK	KY
08694	x		MSNA	AN		08784	x		MSNC	CD
08695	x		MSNC	CD		08786	a	D	FSWK	CF
08696	a	D	MSSW	WN		08788	x		MSNX	DY (S)

08790	x		MSNL	LO	08894	x		MSNA	AN
08792	x		MSSL	LA	08896	x		FSWK	CF
08795	x	D	FSWX	CF (S)	08897	x	D	MSSA	BR
08798	x		FSWK	CF	08899	x		MSNE	DY
08799	x		MSNA	AN	08900	x	D	MSNA	AN
08801	x		MSSL	LA	08901	xn		MSNX	BS (S)
08805	x	F0	MSNB	BS	08902	x		MSNA	AN
08806	a	F	FSCK	KY	08903	x		FSCD	DR
08807	x		MSSB	BY	08904	x		MSSO	OC
08809	x		MSNA	AN	08905	x		MSSR	RG
08810	a		MSSN	NC	08906	x		FSSM	ML
08811	a*		MSSX	SF (S)	08907	x	0	MSNC	CD
08813	a	D	FSCD	DR	08908	xz		FSNL	NL
08815	x		MSNA	AN	08909	x		MSSS	SF
08817	x	BS	MSNA	AN	08910	x		MSNU	CL
08818	x		MSSA	BR	08911	x	D	MSNU	CL
08819	x	D	MSSL	LA	08912	x		MSNU	CL
08823	a		MSCH	CD	08913	x	D	MSNA	AN
08824	a	F	FSCD	DR	08914	x		MSSB	BY
08825	a		MSSO	OC	08915	x	F	MSNL	LO
08826	a		MSNU	CL	08918	x	D	MSNA	AN
08827	a		MSNU	CL	08919	x		FSNX	TI
08828	a		MSSS	SF	08920	x	F	MSNB	BS
08830	x*		FSWK	CF	08921	x		MSNC	CD
08834	x	F	MSSS	SF	08922	x	D	FSSM	ML
08837	x*	D	MSSO	OC	08924	x	D	MSSR	RG
08842	x		MSSW	WN	08925	x		MSNA	AN
08844	x		MSNX	CL (S)	08926	x		MSSW	WN
08847	x*		NKJD	EH	08927	x		MSSB	BY
08853	xr		FSSM	ML	08928	x	FR	MSSN	NC
08854	x*		NKJH	SU	08931	x		FSNH	HT
08856	x		MSNA	AN	08932	x		FSWK	CF
08865	x		MSSC	CA	08933	x*	0	NKJD	EH
08866	x		MSCH	DR	08934	x		MSSW	WN
08867	x	0	FSWX	CF (S)	08937	x	D	MSSL	LA
08869	x	G	MSSN	NC	08938	xr	0	FSSX	ML
08872	x	D	MSNA	AN	08939	x		MSNX	AN
08873	x	M	MSSW	WN	08940	x		NKJD	EH
08877	x	D	FSCD	DR	08941	x		MSSL	LA
08878	x		MDYX	BS (U)	08942	x		FSWK	CF
08879	x		FSNT	TI	08944	x	D	MSSO	OC
08881	x	D	FSSM	ML	08946	x	D	MSSR	RG
08882	x		FSSB	ML	08947	x		MSSO	OC
08883	x	0	FSSM	ML	08948	x		MSSO	OC
08884	x		MSNA	AN	08950	x	I	FSNL	NL
08886	x		FSNH	HT	08951	x	D	MSNA	AN
08887	x		MSSW	WN	08952	x		FSSM	ML
08888	xz	R	FSNH	HT	08953	x	D	MSSL	LA
08890	x	D	MSSW	WN	08954	x	F	MSSL	LA
08891	x		MSNL	LO	08955	x		MSSL	LA
08892	x*	D	NKJD	EH	08956	x		MSNE	DY
08893	x	D	MSNX	BS (S)	08957	x		MSSS	SF

08958 x   MSSS   SF

Names:

08562	The Doncaster Postman		08757	EAGLE C.U.R.C.
08578	Libert Dickinson		08869	The Canary
08633	The Sorter		08888	Postman's Pride
08649	G.H. Stratton		08950	Neville Hill 1st
08701	GATESHEAD TMD 1852 – 1991			

Class 08/9. Fitted with cut-down cab and headlight for Cwmmawr branch.

08993	(08592)	x		FSWL	LE	ASHBURNHAM
08994	(08462)	a	**FR**	FSWL	LE	GWENDRAETH
08995	(08687)	a	**FC**	FSWL	LE	KIDWELLY

CLASS 09   BR SHUNTER   0 – 6 – 0

Built: 1959 – 62 by BR at Darlington or Horwich Works.
Engine: English Electric 6KT of 298 kW (400 hp) at 680 rpm.
Main Generator: English Electric 801.
Traction Motors: English Electric 506.
Max. Tractive Effort: 111 kN (25000 lbf).
Cont. Tractive Effort: 39 kN (8800 lbf) at 11.6 mph.
Power At Rail: 201 kW (269 hp).
Brake Force: 19 t.
Weight: 50 t.
Max. Speed: 27 mph.

Length over Buffers: 8.92 m.
Wheel Diameter: 1372 mm.
RA: 5.

Class 09/0 were formerly numbered 3665 – 71, 3719 – 21, 4099 – 4114.

CLASS 09/0. Built as Class 09.

09001		FSWK	CF		09014	**D**	FSCK	KY
09003		NKJH	SU		09015	**D**	FSWK	CF
09004		NKJH	SU		09016	**D**	NKJH	SU
09005	**D**	FSCK	KY		09018		NKJH	SU
09006		NKJH	SU		09019	**D**	NKJH	SU
09007		NKJH	SU		09020		NKJH	SU
09008	**D**	FSWK	CF		09021		NKJH	SU
09009	**D**	NKJH	SU		09022		NKJH	SU
09010	**D**	NKJH	SU		09023		NKJH	SU
09011	**D**	NKJH	SU		09024	**D**	NKJH	SU
09012	**D**	NKJH	SU		09025		NKJH	SU
09013	**D**	FSWK	CF		09026	**D**	NKJH	SU

Names:

09009	Three Bridges C.E.D.	09026	William Pearson
09012	Dick Hardy		

CLASS 09/1. Converted from Class 08. 90 V electrical equipment.

09101	(08833)	**D**	MSSR	RG
09102	(08832)	**D**	MSSR	RG
09103	(08766)	**D**	FSSM	ML

09104	(08749)	**D**	FSCN	TO
09105	(08835)	**D**	FSWK	CF
09106	(08759)	**D**	FSNY	TE
09107	(08845)	**D**	FSWK	CF

CLASS 09/2. Converted from Class 08. 100 V electrical equipment.

09201	(08421)	**D**	FSCN	TO
09202	(08732)	**D**	FSSM	ML
09203	(08781)	**D**	FSWK	CF
09204	(08717)	**D**	FSNY	TE
09205	(08620)	**D**	FSSB	ML

CLASS 20   ENGLISH ELECTRIC TYPE 1   Bo – Bo

Built: 1957 – 68 by English Electric Company at Vulcan Foundry, Newton le Willows or Robert Stephenson & Hawthorn, Darlington. 20001 – 128 were originally built with disc indicators whilst 20129 – 228 were built with four character headcode panels.
Engine: English Electric 8SVT Mk. II of 746 kW (1000 hp) at 850 rpm.
Main Generator: English Electric 819/3C.
Traction Motors: English Electric 526/5D (20001 – 48) or 526/8D (others).
Max. Tractive Effort: 187 kN (42000 lbf).
Cont. Tractive Effort: 111 kN (25000 lbf) at 11 mph.
Power At Rail: 574 kW (770 hp).   **Length over Buffers:** 14.25 m.
Brake Force: 35 t.   **Wheel Diameter:** 1092 mm.
Design Speed: 75 mph.   **Weight:** 73.5 t.
Max. Speed: 60 mph.   **RA:** 5.
Train Brakes: Air & vacuum.
Multiple Working: Blue Star Coupling Code.

Formerly numbered in series 8007 – 8190, 8315 – 8325.

CLASS 20/0. BR-owned Locomotives.

20007	st		TAKX	KI (S)	20118	**FR**	CDJB	BS
20016	st		FCXX	TO (S)	20121	st	TAKX	KI (U)
20032	s		TAKX	KI (S)	20128	st	TAKB	BS
20057	st		FCXX	TO (S)	20131	st **T**	TAKB	BS
20059	st **FR**		FCXX	TO (S)	20132	st **FR**	CDJB	BS
20066			CEJX	BS (S)	20137	**FR**	FMXX	TE (U)
20072	st		TAKX	KI (U)	20138	**FR**	CEJX	BS (S)
20073	st		FCXX	SP (U)	20154	st	FCXX	TO (U)
20075	st		TAKB	BS	20165	**FR**	CDJB	BS
20081	st		FCXX	TO (S)	20168	st	FCXX	TO (S)
20087	st		CDJB	BS	20169	st **CS**	CDJB	BS
20092	**CS**		CDJB	BS	20187	st	TAKB	BS
20104	st **FR**		TAKX	KI (S)	20190	st	TAKX	KI (U)
20117	st		TAKX	KI (U)	20215	st **FR**	TAKX	KI (U)

CLASS 20/9. Privately-owned by Hunslet – Barclay Ltd.

Used on summer weedkilling trains. Stored at Kilmarnock during the winter.
Non-standard Liveries: 20901 – 6 are in Hunslet – Barclay two-tone grey livery with red lettering.

20901	(20041)	t	**0**	XYPD	HB	NANCY
20902	(20060)		**0**	XYPD	HB	LORNA
20903	(20083)		**0**	XYPD	HB	ALISON
20904	(20101)		**0**	XYPD	HB	JANIS
20905	(20225)	t	**0**	XYPD	HB	IONA
20906	(20219)		**0**	XYPD	HB	Kilmarnock 400

CLASS 31   BRUSH TYPE 2   A1A – A1A

Built: 1957 – 62 by Brush Traction at Loughborough.
31102/5 – 7/10/25/34/44/418/50/61/544 retain two headcode lights. Others have roof-mounted headcode boxes.
Engine: English Electric 12SVT of 1100 kW (1470 hp) at 850 rpm.
Main Generator: Brush TG160-48.
Traction Motors: Brush TM73-68.
Max. Tractive Effort: 160 kN (35900 lbf) (190 kN (42800 lbf)*).
Cont. Tractive Effort: 83 kN (18700 lbf) at 23.5 mph. (99 kN (22250 lbf) at 19.7 mph *.)
Power At Rail: 872 kW (1170 hp).   **Length over Buffers:** 17.30 m.
Brake Force: 49 t.   **Driving Wheel Diameter:** 1092 mm.
Design Speed: 90 (80*) mph.   **Centre Wheel Diameter:** 1003 mm.
Max. Speed: 60 mph (90 mph 31/4).   **Weight:** 107 – 111 t.
RA: 5 or 6.   **ETH Index (Class 31/4):** 66.
Train Brakes: Air & vacuum.
Multiple Working: Blue Star Coupling Code.
Communication Equipment: This class is in the process of being fitted with cab to shore radio-telephone.

Non standard liveries:

31116 is red, yellow, red and grey with 'Infrastructure' branding.
31413 is BR blue with yellow cabsides, a light blue stripe along the bottom of the body and a red band around the bottom of the cabs.

Formerly numbered 5520 – 5699, 5800 – 5862 (not in order).

CLASS 31/1. Standard Design. RA5.

31102	**C**	CEJB	BS		31142	**C**	IWJC	CD
31105	* **C**	CEJB	BS		31144	**C**	IWJC	CD
31106	* **C**	CEJB	BS		31145	**C**	IMJW	BS
31107	**C**	CEJB	BS		31146 r	**C**	RDDJ	BS
31110	**C**	CEJB	BS		31147 r	**C**	RDDJ	BS
31112	* **C**	CEJB	BS	◌	31149	**FR**	IEJW	IM
31113	**C**	CEJB	BS		31154	**C**	IWJC	CD
31116	**0**	IMJB	BS		31155	**FA**	RDJB	BS
31119	**C**	IMJW	BS		31158	**C**	RDDJ	BS
31125	**C**	CEJX	BS		31159	**C**	IWJC	CD
31126	**C**	IMJW	BS		31160	**F**	RCWC	CD
31128	**FO**	RDJW	BS		31163	**C**	IWJC	CD
31130	**FC**	FCFN	TO		31164	**FO**	RDJW	BS
31132	**FO**	RDJW	BS	◌	31165	**G**	NKJW	SF
31134	**C**	IMJW	BS		31166	**C**	RDDJ	BS
31135	**C**	NKJS	SF		31171	**FO**	IXXS	BS (U)

31174	C	RDJB	BS	31242	C	RCJC	CD
31178	C	RDJB	BS	31247	FR	RBJW	IM
31180	FR	NKJW	SF	31248	FO	IXXS	BS (U)
31181	C	NKJS	SF	31250	C	NKJS	SF
31184	FO	IEJW	IM	31252	FO	IMJW	BS
31185	C	RDJB	BS	31255	C	RCJC	CD
31186	C	RDJM	SF	31263	C	RCWC	CD
31187	C	RDJS	SF	31268	C	NKJW	SF
31188	C	RCJC	CD	31270	FC	RCWC	CD
31190	C	RCWC	CD	31271	FA	IHFB	BS
31191	C	NKJW	SF	31272	C	RCJC	CD
31199	FC	FCFN	TO	31273	C	RDJB	BS
31200	FC	FCFN	TO	31275	FC	FCFN	TO
31201	FC	FCFN	TO	31276	FC	IEJW	IM
31203	C	IWJC	CD	31282	FR	RCWC	CD
31205	FR	IEJW	IM	31285	C	RCJC	CD
31206	C	IWJI	BS	31290	C	NKJS	SF
31207	C	RCJC	CD	31294	FA	IEJW	IM
31209	FA	IMJW	BS	31301	FR	RCWC	CD
31219	C	NKJW	SF	31302	FP	FCFN	TO
31224	C	FCFN	TO	31304	FC	FCFN	TO
31229	C	RCJC	CD	31306	C	RCJC	CD
31230 *	FO	RBJW	IM	31308	C	RDJB	BS
31232	C	IWJI	BS	31312	FC	FCFN	TO
31233	C	RCJC	CD	31317	FO	RDJW	BS
31235	C	IWJC	CD	31319	FC	FCFN	TO
31237	C	RDJB	BS	31327	FR	RCWC	CD
31238	C	RCJC	CD				

Names:

31102	Cricklewood	31130	Calder Hall Power Station
31105	Bescot TMD	31146	Brush Veteran
31106	The Blackcountryman	31147	Floreat Salopia
31107	John H Carless VC	31165	Stratford Major Depot
31116	RAIL Celebrity	31233	Severn Valley Railway

CLASS 31/4. Equipped with Train Heating. RA6.
CLASS 31/5. Dedicated for Civil Engineer's Department Use. Train Heating Equipment isolated. RA6.

31403		IHFB	BS	
31405	M	CEJB	BS	Mappa Mundi
31407	(31507)	M	IHFB	BS
31408		RCKC	CD	
31410	RR	RCKC	CD	Granada Telethon
31411	(31511)	D	IMJW	BS
31512	(31412)	C	IWJC	CD
31413		O	IXXS	BS (U)
31514	(31414)	C	IWJI	BS
31415			CEJB	BS
31516	(31416)	C	IWJB	BS
31417		D	IWJB	BS
31418			IMJW	BS

31519 (31419)	**C**	IWJC	CD	
31420 (31172)	**M**	IWJB	BS	
31421 (31140)	**RR**	RCKC	CD	Wigan Pier
31422 (31522)	**M**	IWJB	BS	
31423 (31197)	**M**	IWJB	BS	Jerome K. Jerome
31524 (31424)	**C**	IWJB	BS	
31526 (31426)	**C**	IWJI	BS	
31427 (31194)		IMJW	BS	
31530 (31430)	**C**	IWJB	BS	Sister Dora
31531 (31431)	**C**	IEJI	IM	
31432 (31153)		RCKC	CD	
31533 (31433)	**C**	IWJI	BS	
31434 (31258)		IWJB	BS	
31435 (31179)	**C**	IHFB	BS	
31537 (31437)	**C**	IWJB	BS	
31438 (31139)		RCWC	CD	
31439 (31239)	**RR**	RCKC	CD	North Yorkshire Moors Railway
31541 (31441)	**C**	IEJI	IM	
31544 (31444)	**C**	IEJI	IM	Keighley and Worth Valley Railway
31545 (31445)		IWJB	BS	
31546 (31446)	**C**	IWJB	BS	
31547 (31447)	**C**	IEJW	IM	
31548 (31448)	**C**	IWJI	BS	
31549 (31449)	**C**	IEJI	IM	
31450 (31133)		IWJB	BS (U)	
31551 (31451)	**C**	IWJB	BS	
31552 (31452)	**C**	IEJI	IM	
31553 (31453)	**C**	IEJW	IM	
31554 (31454)	**C**	IWJB	BS	
31555 (31455)	**RR**	RCKC	CD	'Our Eli'
31556 (31456)	**C**	IEJI	IM	
31457 (31169)	**D**	IWJB	BS	
31558 (31458)	**C**	IEJI	IM	
31459 (31256)		IHFB	BS	
31461 (31129)	**D**	IHFB	BS	
31462 (31315)	**D**	CEJB	BS	
31563 (31463)	**C**	RBJW	IM	
31465 (31565)	**RR**	RCKC	CD	
31466 (31115)	**C**	IHFB	BS	
31467 (31216)		CEJB	BS	
31468 (31568)	**C**	RDJB	BS	The Enginemans Fund
31569 (31469)	**C**	RDJW	BS	

CLASS 33   BRCW TYPE 3   Bo – Bo

Built: 1960 – 62 by the Birmingham Railway Carriage & Wagon Company, Smethwick.
Engine. Sulzer 8LDA28 of 1160 kW (1550 hp) at 750 rpm.
Main Generator: Crompton Parkinson CG391B1.
Traction Motors: Crompton Parkinson C171C2.
Max. Tractive Effort: 200 kN (45000 lbf).

Cont. Tractive Effort: 116 kN (26000 lbf) at 17.5 mph.
Power At Rail: 906 kW (1215 hp). **Length over Buffers:** 15.47 m.
Brake Force: 35 t. **Wheel Diameter:** 1092 mm.
Design Speed: 85 mph. **Weight:** 77.5 t (78.5 t Class 33/1).
Max. Speed: 60 mph. **RA:** 6.
Train Heating: Electric (y isolated). **ETH Index:** 48.
Train Brakes: Air & vacuum.
Multiple Working: Blue Star Coupling Code.
Communication Equipment: This class is in the process of being fitted with cab
to shore radio-telephone.

Formerly numbered in series 6500 – 97 but not in order.

Class 33/0. Standard Design.

33002	y	C	NKJM	SL	Sea King
33008	y	G	NKJE	EHSL Eastleigh
33012	e		NKJR	SL
33019	e	C	NKJE	EHSL Griffon
33021	e	FA	NKJR	SL
33023	e		NKJR	SL
33025	e	C	NKJE	EH SL
33026	e	C	NKJR	SL	Seafire
33030	e	C	NKJE	EH S L
33035	y	N	NKJE	EH S L
33042	ys	FA	NKJR	SL
33046	y	C	NKJE	EHSL Merlin
33048	es		NKJR	SL
33051	e	C	NKJE	EHSL Shakespeare Cliff
33052	e		NKJR	SL	Ashford
33053	e	FA	NKJR	SL
33057	ys	C	NKJM	SL	Seagull
33063	ys	FA	NKJR	SL
33064	e	FA	NKJM	SL
33065	e	C	NKJR	SL	Sealion

**Class 33/1. Fitted with Buckeye Couplings & SR Multiple Working
Equipment for use with SR EMUs, TC stock & class 73.**

Also fitted with flashing light adaptor for use on Weymouth Quay line.

33103				
33109	e	D	NKJM	SL	Captain Bill Smith RNR
33116	e	D	NKJE	EH SL	Hertfordshire Rail Tours
33117				

Class 33/2. Built to Former Loading Gauge of Hastings Line.

33202	ys	C	NKJM	SL	The Burma Star
33204	es	FD	NXJX	SL (S)
33206	es	FD	NXJX	SL (S)
33207	ys	FA	NKJR	SL	Earl Mountbatten of Burma
33208	ys	C	NKJM	SL

CLASS 37   ENGLISH ELECTRIC TYPE 3   Co-Co

Built: 1960 – 5 by English Electric Company at Vulcan Foundry, Newton le
Willows or Robert Stephenson & Hawthorn, Darlington. 37003 – 116/350/1/9

Trainload Construction

Trainload Coal

Trainload Metals

Trainload Petroleum

Railfreight Distribution

Trainload Freight and Railfreight Distribution sub-sector markings as used on locomotives.

Class 03 No. 03079 is pictured in faded BR blue livery at Sandown, IoW on 9th October 1993. *C.J. Marsden*

▲ Departmental grey liveried Class 08 No. 08922 shunts an engineers tra at Perth during July 1993. *Norman Barringto*

▼ Class 09 No. 09013 at Coalville on 3rd June 1990. *A.O. Wyn*

Class 20s Nos. 20066 & 20138, the latter in Railfreight red-stripe livery, approach Droitwich with a Worcester open day special on 2nd May 1993.
P.J. Chancellor

Regional Railways liveried Class 31 No. 31465 accelerates away from Grange-over-sands on 22nd May 1993 with the 09.30 Barrow – Preston.
Dave McAlone

Class 31 No. 31554 in Civil Engineers livery passes Washwood Heath with a westbound engineers train on

Class 33 No. 33035 in Network SouthEast livery at Marcroft sidings, Eastleigh on 29th June 1993.

Brian Denton

▲ BR blue Class 37 No. 37275 'Oor Wullie' stands outside BRML Glasgow on 15th October 1993 following a repaint. *Brian Morrison*

▼ Class 37 No. 37416 in Mainline livery at Lostwithiel with a china clay train bound for Fowey docks on 2nd August 1993. *Stephen Widdowson*

Trainload Petroleum liveried Class 37 No. 37717 'Stainless Pioneer' at Dalston with a train of empty tanks for Grangemouth on 23rd September 1993.

Kevin Conkey

Class 43 No. 43027 leads the 14.48 Penzance – Paddington at Aller Junction on 7th May 1993.

with the exception of 37019*/031/047/053/065*/072*/073/074/075*/100*
(* one end only) retain box-type route indicators, the remainder having central
headcode panels/marker lamps.
Engine: English Electric 12CSVT of 1300 kW (1750 hp) at 850 rpm.
Main Generator: English Electric 822/10G.
Traction Motors: English Electric 538/A.
Max. Tractive Effort: 245 kN (55500 lbf).
Cont. Tractive Effort: 156 kN (35000 lbf) at 13.6 mph.
Power At Rail: 932 kW (1250 hp).   **Length over Buffers:** 18.75 m.
Brake Force: 50 t.   **Wheel Diameter:** 1092 mm.
Design Speed: 90 mph.   **Weight:** 103 – 108 t.
Max. Speed: 80 mph.   **RA:** 5 or 7.
Train Heating: Electric (Class 37/4 only). **ETH Index:** 38
Train Brakes: Air & vacuum.
Multiple Working: Blue Star Coupling Code.
Communication Equipment: This class is in the process of being fitted with cab
to shore radio-telephone.

a Vacuum brake isolated.

Formerly numbered 6600 – 8, 6700 – 6999 (not in order). 37271 – 4 are the
second locos to carry these numbers. They were renumbered to avoid confu-
sion with Class 37/3 locos.

Class 37/0. Unrefurbished Locos. Technical details as above. RA5.

37003	+	C	IEJI	IM
37004		FM	FQXA	ML (U)
37009	+	FD	FPJW	IM
37010		C	IGJK	BR
37012		C	IGJK	BR
37013	+	F	NKJS	SF
37015	+	FD	MDTT	TI
37019	+	FD	MDTT	TI
37023		C	NKJS	SF	Stratford
37025		C	RAJV	IS
37026 (37320)	+	FD	MDTT	TI	Shapfell
37029	+	FD	MDYX	CE (U)
37031	+	FD	FQXA	CF
37032 (37353)	+	FR	MDYX	TI (U)
37035		C	IGJK	BR
37037 (37321)		FM	FQXA	CF
37038		C	IGJK	BR
37040		FM	IGJA	BR
37042	+	FM	IGJA	BR
37043 (37354)		C	RAJV	IS
37045 (37355)	+	F	RBJN	HT
37046		C	FQXA	TO
37047	+	FD	NKJS	SF
37048		FM	IGJA	BR
37049		C	RBJI	IM	Imperial
37051		FM	FCPM	ML SL
37053	+	FD	MDTT	TI
37054		C	IGJK	BR

+	37055	+	**FD** NKJS	SF	
⌒	37057	+	**BR** MDSR	TI (S)	
	37058	+	**C** IEJI	IM	
	37059	+	**FD** RBJN	HT	Port of Tilbury
	37063	+	**FD** RBJN	HT	
⌒	37065	+	**FD** IGJA	BR	
	37066	+	**C** FCPM	ML	
	37068 (37356)	+	**FD** MDTT	TI	Grainflow
	37069	+	**C** RAJV	IS	
⌒	37070		**FD** MDYX	TI (U)	
	37071	+	**C** FCPM	ML	
⌒	37072	+	**D** IGJA	BR	
	37073	+	**FD** MDTT	TI	Fort William/An Gearasdan
⌒	37074	+	**FD** IGJA	~~BR~~ SL	
	37075	+	**F** MDTT	TI	
+	37077		**FM** IGJA	~~BR~~ SL	
	37078		**FM** IISA	IS	
	37079 (37357)	+	**FD** MDTT	TI	Medite
	37080		**FP** IISA	IS	
	37083		**C** RBJI	IM	
	37087		**C** RAJV	IS	
	37088 (37323)		**C** RAJV	IS	Clydesdale
⌒	37092		**C** FQXA	TO	
	37095	+	**C** IEJI	IM	
⌒	37097		**C** IGJK	BR	
⌒	37098		**C** IGJK	BR	
	37099 (37324)		**C** RAJV	IS	Clydebridge
	37100	+	**FM** FCPM	ML	
	37101	+	**FD** FQXA	CF	
	37104		**C** IEJI	IM	
+	37106	+	**C** RDJS	SF	
	37107	+	**FD** MDTT	TI	
	37108 (37325)	+	**F** MDTT	TI	
+	37109		**FM** IGJA	~BRSL~	
	37110	+	**FD** MDTT	TI	
	37111 (37326)		**FM** FCPM	ML	Glengarnock
	37113	+	**FD** IISA	IS	Radio Highland
⌒	37114	+	**C** RDKB	BS	City of Worcester
	37116	+	**BR** FCPM	ML	
	37128	+	**BR** RBJN	HT	
	37131	+	**FD** MDTT	TI	
	37133		**C** IISA	IS	
⌒	37137 (37312)		**FM** IGJA	BR	Clyde Iron
⌒	37138		**FM** IGJA	BR	
	37139	+	**FC** RBJH	HT	
+	37140		**C** NKJS	SF	
	37141		**C** REJK	CF	
	37142		**C** REJK	CF	
	37144	r	**FA** RBJI	IM	
	37146		**C** REJK	CF	
	37152 (37310)		**I** IISA	IS	
	37153		**C** RAJV	IS	

Number						
37154		+	FD	MDTT	TI	Johnson Stevens Agencies
37156	(37311)	r	C	RAJV	IS	British Steel Hunterston
37158			C	REJK	CF	
37162		+	D	RDKB	BS	
37165	(37374)	+	C	RAJV	IS	
37167		+	FC	RDJM	SF SL	
37170		r	C	IISA	IS	
37174			C	IGJK	BR	
37175			C	IISA	IS	
37178		+	FD	MDTT	TI	
37184			C	FCPM	ML	
37185		+	C	RDKB	BS	Lea & Perrins
37188			C	FCPM	ML	
37191			C	REJK	CF	
37194		+	FD	NKFE	EH SL	British International Freight Association
37196			C	RAJV	IS	
37197		+	C	REJS	CF	
37198		+	C	NKJE	EH SL	
37201			C	RAJV	IS	Saint Margaret
37202			FM	RBJN	HT	
37203			FM	IGJA	BR	
37207			C	REJK	CF	
37209			BR	MDYX	TI (U)	
37211			C	RAJV	IS	
37212		+	FC	FCPM	ML	
37213		+	FC	IGJA	BR	
37214		+	FA	IISA	IS	
37216		r +	G	RDJS	SF	Great Eastern
37217		+		RBJN	HT	
37218		+	FD	MDTT	TI	
37219		r		IGJA	BR SL	
37220		+	FP	NKFE	EH SL	
37221			I	IISA	IS	
37222		+	FC	IGJA	BR	
37223		+	FC	FQXA	CF	
37225		+	FD	MDTT	TI	
37227		+	FM	IGJA	BR	
37229		+	FC	FQXA	CF	
37230		+	C	REJS	CF	
37232		r	C	RAJV	IS	The Institution of Railway Signal Engineers
37235		+	F	FQXA	SL	
37238		+	FD	MDTT	TI	
37239		+	FC	MDSR	TI (U)	The Coal Merchants' Association of Scotland
37240		+	C	RAJV	IS	
37241			FM	NKJW	SF	
37242		+	FD	NKJW	SF	
37244		+	FD	NKJS	SF	
37245			C	NKFE	EH SL	
37248		+	FM	MDYX	TI (U)	

37250		+	**FM** IISA	IS	
37251		+	**I** IISA	IS	The Northern Lights
37252			**FD** MDYX	TI (U)	
37254		+	**C** REJS	CF	
37255		+	**C** RAJV	IS	
37258		+	**C** REJS	CF	
37261		+	**FD** MDTT	TI	Caithness
37262		+	**D** IISA	IS	Dounreay
37263			**C** REJK	CF	
37264			**C** IGJK	BR	
37271	(37303)	+	**FD** IEJW	IM	
37272	(37304)	+	**FD** RBJN	HT	
37274	(37308)	+	**C** NKJE	EH SL	
37275		+	**R**AJV	IS	Oor Wullie
37278		+	**FC** MDSR	TI (U)	
37280		+	**FP** MDSR	TI (U)	
37285		+	**F** RBJN	HT	
37293		+	**FM** NKJE	EH SL	
37294		+	**C** RAJV	IS	
37298		+	**FD** MDTT	TI	

Class 37/3. Unrefurbished locos fitted with regeared (CP7) bogies.
Details as Class 37/0 except:
Max. Tractive Effort: 250 kN (56180 lbf).
Cont. Tractive Effort: 184 kN (41250 lbf) at 11.4 mph.

37350	(37119)	+	**FP** FPRI	IM	
37351	(37002)	+	**C** RAJV	IS	
37358	(37091)		**F** FPRI	IM	P & O Containers
37359	(37118)		**FP** FPRI	IM	
37370	(37127)		**C** NKJS	SF	
37371	(37147)	+	**C** NKJS	SF	
37372	(37159)		**C** IGJK	BR SL	
37373	(37160)		**FR** MDSR	TI (U)	
37375	(37193)	+	**C** NKFE	EH SL	
37376	(37199)	+	**FC** NKJS	SF	
37377	(37200)	+	**C** NKJE	EH SL	
37378	(37204)	+	**FD** FPRI	IM	
37379	(37226)		**C** NKJS	SF	
37380	(37259)		**FC** NKFE	EH SL	
37381	(37284)	+	**FD** FPYX	IM (U)	
37382	(37145)		**FP** FPYI	IM	

Class 37/4. Refurbished locos fitted with train heating. Main generator replaced by alternator. Regeared (CP7) bogies. Details as class 37/0 except:

Main Alternator: Brush BA1005A.
Max. Tractive Effort: 256 kN (57440 lbf).
Cont. Tractive Effort: 184 kN (41250 lbf) at 11.4 mph.
Power At Rail: 935 kW (1254 hp).
All have twin fuel tanks.

37401	(37268)	r	**FD** MDRM	ML	Mary Queen of Scots
37402	(37274)	r	**M** RCMC	CD	

37403	(37307) r	**FD** FCPM	ML	Glendarroch
37404	(37286) r	**M** FCPM	ML	
37405	(37282) r	**M** FABI	IM	Strathclyde Region
37406	(37295) r	**FD** MDRM	ML	The Saltire Society
37407	(37305) r	**M** RCMC	CD	Loch Long
37408	(37289)	**BR** RCMC	CD	Loch Rannoch
37409	(37270) r	**M** MDRM	ML	Loch Awe
37410	(37273) r	**M** MDRM	ML	Aluminium 100
37411	(37290)	**FD** MDRL	LA	
37412	(37301)	**FD** MDRL	LA	
37413	(37276) r	**FD** MDRL	LA	Loch Eil Outward Bound
37414	(37287) r	**RR** RCMC	CD	Cathays C&W Works 1846 – 1993
37415	(37277) r	**M** FMPY	TE	
37416	(37302) r	**M** MDRL	LA	
37417	(37269) r	**M** FABI	IM	Highland Region
37418	(37271) r	**FP** RCMC	CD	
37419	(37291) r	**M** FMPY	TE	
37420	(37297) r	**M** FABI	IM	The Scottish Hosteller
37421	(37267) r	**RR** RCMC	CD	The Kingsman
37422	(37266) r	**RR** RCMC	CD	Robert F. Fairlie Locomotive Engineer 1831-1885
37423	(37296) r	**M** MDRM	ML	Sir Murray Morrison

37423 1873 – 1948 Pioneer of British Aluminium Industry

37424	(37279) r	**M** MDRM	ML	Isle of Mull
37425	(37292) r	**FA** RCMC	CD	Sir Robert McAlpine/ Concrete Bob (opp. sides)
37426	(37299) r	**M** FMPY	TE	
37427	(37288) r	**RR** RAJP	IS	Highland Enterprise
37428	(37281) r	**FP** RAJP	IS	David Lloyd George
37429	(37300) r	**RR** RCMC	CD	Eisteddfod Genedlaethol
37430	(37265) r	**M** MDRM	ML	Cwmbrân
37431	(37272) r	**M** RAJP	IS	

Class 37/5. Refurbished locos. Main generator replaced by alternator. Regeared (CP7) bogies. Details as class 37/4 except:

Max. Tractive Effort: 248 kN (55590 lbf).
All have twin fuel tanks.

37501	(37005)	**FM** FPJI	IM	
37502	(37082)	**FM** FPJW	IM	
37503	(37017)	**FM** RCLC	CD	
37504	(37039)	**FM** RCLC	CD	
37505	(37028)	**I** IISA	IS	
37506	(37007)	**FM** FPTY	TE	British Steel Skinningrove
37507	(37036)	**FM** FPYX	IM (U)	
37508	(37090) s	**FM** FPRI	IM	
37509	(37093)	**FM** RCLC	CD	
37510	(37112)	**I** IISA	IS	
37511	(37103)	**FM** FPYI	IM	Stockton Haulage
37512	(37022)	**FM** FPJW	IM	Thornaby Demon
37513	(37056)	**FM** FPJI	IM	
37514	(37115) s	**FM** FMPY	TE	
37515	(37064) s	**FM** FPJI	IM	

37516 (37086)	s	**FM** FMPY	TE	
37517 (37018)	as	**FM** FPJI	IM	
37518 (37076)		**FM** FQXA	IM	(U)
37519 (37027)		**FM** FPJI	IM	
37520 (37041)		**FM** FABI	IM	
37521 (37117)		**FP** FPEK	CF	
✚ 37667 (37151)	as	**FP** FPFR	IM	
37668 (37257)	s	**FP** FPEK	CF	
37669 (37129)		**FD** MDRL	LA	
37670 (37182)		**FD** MDRL	LA	St. Blazey T&RS Depot
37671 (37247)		**FD** MDRL	LA	Tre Pol and Pen
37672 (37189)	s	**FD** MDRL	LA	Freight Transport Association
37673 (37132)		**FD** MDRL	LA	
37674 (37169)		**FD** MDRL	LA	
37675 (37164)	s	**FD** MDRL	LA	William Cookworthy
✚ 37676 (37126)		**FA** FPFR	IM	
37677 (37121)		**FA** FABI	IM	
✚ 37678 (37256)		**FA** FPFR	IM	
✚ 37679 (37123)		**FA** FPFR	IM	
37680 (37224)		**FA** FPRI	IM	
37682 (37236)		**FA** FMPY	TE	
37683 (37187)		**I** IISA	IS	
37684 (37134)		**FA** FPRI	IM	Peak National Park
37685 (37234)		**FR** IISA	IS	
37686 (37172)		**FA** FABI	IM	
37687 (37181)		**FA** FPJW	IM	
37688 (37205)		**FA** FPCI	IM	Great Rocks
37689 (37195)	s	**FC** FPRI	IM	
37690 (37171)		**FO** FCPM	ML	
37691 (37179)	s	**FO** FPRI	IM	
37692 (37122)	s	**FC** FCPM	ML	The Lass O' Ballochmyle
37693 (37210)	s	**FC** FCPM	ML	
37694 (37192)	s	**FC** FPJI	IM	
37695 (37157)	s	**FC** FCPM	ML	
37696 (37228)	s	**FC** FCPM	ML	
37697 (37243)	s	**FC** FPTY	TE	
37698 (37246)	s	**FC** FPCI	IM	
37699 (37253)		**FC** FPCI	IM	

**Class 37/7. Refurbished locos. Main generator replaced by alternator. Regeared (CP7) bogies. Ballast weights added.

Details as class 37/4 except:
Main Alternator: GEC G564AZ (37796 – 803) Brush BA1005A (others).
Max. Tractive Effort: 276 kN (62000 lbf).
Weight: 120 t.   **RA:** 7.
All have twin fuel tanks.

37701 (37030)	s	**FC** FCKK	CF	
37702 (37020)	s	**FC** FCKK	CF	Taff Merthyr
37703 (37067)	s	**FC** FCKK	CF SL	
37704 (37034)	s	**FC** FCKK	CF	
37705 (37060)	a	**FP** FPFR	IM SL	

37706	(37016)	a	**FP** FPCI	IM	Conidae
37707	(37001)	a	**FP** FPCI	IM	
37708	(37089)	a	**FP** FPCI	IM	
37709	(37014)	a	**FP** FPFR	~~IM~~ SL	
37710	(37044)		**FM** FPFR	IM	
37711	(37085)		**FM** FPCI	IM	
37712	(37102)		**FP** FPGM	ML	Teesside Steelmaster
37713	(37052)		**FM** FPCI	IM	British Steel Workington
37714	(37024)		**FM** FCPM	ML	
37715	(37021)		**FP** FPCI	~~IM~~ SL	British Petroleum
37716	(37094)		**FM** FPTY	TE	British Steel Corby
37717	(37050)		**FP** FPCI	IM	Stainless Pioneer
37718	(37084)		**FM** FPTY	TE	Hartlepool Pipe Mill
37719	(37033)	a	**FP** FPCI	IM	
37796	(37105)	s	**FC** FCKK	CF	
37797	(37081)	s	**FC** FCKK	CF	
37798	(37006)	s	**FC** FPCI	~~IM~~ SL	
37799	(37061)	s	**FC** FCKK	CF	Sir Dyfed/County of Dyfed
37800	(37143)	s	**FC** FPCI	~~IM~~ SL	Glo Cymru
37801	(37173)	s	**FC** FPGM	ML	
37802	(37163)	s	**FC** FCKK	CF	
37803	(37208)	s	**FC** FPCI	~~IM~~ SL	
37883	(37176)		**FM** FPCI	IM	
37884	(37183)		**F** FPCI	IM	Gartcosh
37885	(37177)		**FM** FPCI	IM	
37886	(37180)		**FM** FPCI	IM	
37887	(37120)	s	**FC** FCKK	CF	Caerphilly Castle
					Castell Caerffili (opp. sides)
37888	(37135)		**FP** FPYX	IM (U) Petrolea	
37889	(37233)		**FC** FCKK	CF	
37890	(37168)	a	**FP** FPFR	~~IM~~ SL	
37891	(37166)		**FP** FPCI	~~IM~~ SL	
37892	(37149)		**FP** FPFR	~~IM~~ SL	
37893	(37237)		**FP** FPGM	ML	
37894	(37124)	s	**FC** FCKK	CF	
37895	(37283)	s	**FC** FCKK	CF	
37896	(37231)	s	**FC** FCKK	CF	
37897	(37155)	s	**FC** FCKK	CF	
37898	(37186)	s	**FC** FCKK	CF	Cwmbargoed DP
37899	(37161)	s	**FC** FCKK	CF	County of West Glamorgan/
					Sir Gorllewin Morgannwg

Class 37/9. Refurbished Locos. Fitted with manufacturers prototype power units and ballast weights. Main generator replaced by alternator. Details as class 37/0 except:
Engine: Mirrlees MB275T of 1340 kW (1800 hp) at 1000 rpm (37901 – 4), Ruston RK270T of 1340 kW (1800 hp) at 900 rpm (37905 – 6).
Main Alternator: Brush BA1005A (GEC G564, 37905/6).
Max. Tractive Effort: 279 kN (62680 lbf).
Cont. Tractive Effort: 184 kN (41250 lbf) at 11.4 mph.
Weight: 120 t.   **RA:** 7.

All have twin fuel tanks.

37901	(37150)		**FM** FMHK	CF	Mirrlees Pioneer
37902	(37148)		**FM** FMHK	CF	
37903	(37249)		**FM** FMHK	CF	
37904	(37125)		**FM** FMHK	CF	
37905	(37136)	s	**FM** FMHK	CF	Vulcan Enterprise
37906	(37206)	s	**FM** FMHK	CF	

CLASS 43   HST POWER CAR   Bo – Bo

Built: 1976 – 82 by BREL Crewe Works. Formerly numbered as coaching stock but now classified as locomotives. Include luggage compartment.
Engine: Paxman Valenta 12RP200L of 1680 kW (2250 hp) at 1500 rpm. (Mirrlees MB190 of 1680 kW (2250 hp)*.
Main Alternator: Brush BA1001B.
Traction Motors: Brush TMH68 – 46 or GEC G417AZ (43124 – 151/180). Frame mounted.
Max. Tractive Effort: 80 kN (17980 lbf).
Cont. Tractive Effort: 46 kN (10340 lbf) at 64.5 mph.
Power At Rail: 1320 kW (1770 hp). **ETH:** Non standard 3-phase system.
Brake Force: **Length over Buffers:** 17.79 m.
Weight: 70 t. **Wheel Diameter:** 1020 mm.
Max. Speed: 125 mph. **RA:** 5.
Train Brakes: Air.
Multiple Working: With one other similar vehicle.
Communication Equipment: All equipped with driver – guard telephone and cab to shore radio-telephone.

§ Modified to be able to remotely control a class 91 locomotive and to be remotely controlled by a class 91 locomotive.

43002	I	IWRP	PM	Top of the Pops
43003	I	IWRP	PM	
43004	I	IWRP	PM	Swan Hunter
43005	I	IWRP	PM	
43006	I	IWRP	LA	
43007	I	IWRP	LA	
43008	I	IWRP	LA	
43009	I	IWRP	PM	
43010	I	IWRP	PM	
43011	I	IWRP	PM	Reader 125
43012	I	IWRP	PM	
43013 §	I	ICCS	EC	
43014 §	I	ICCS	EC	
43015	I	IWRP	LA	
43016	I	IWRP	PM	Gwyl Gerddi Cymru 1992
				Garden Festival Wales 1992
43017	I	IWRP	PM	
43018	I	IWRP	PM	
43019	I	IWRP	PM	Dinas Abertawe/City of Swansea
43020	I	IWRP	LA	John Grooms
43021	I	IWRP	LA	
43022	I	IWRP	LA	
43023	I	IWRP	LA	County of Cornwall

43024	I	IWRP	LA	
43025	I	IWRP	LA	
43026	I	IWRP	LA	City of Westminster
43027	I	IWRP	LA	
43028	I	IWRP	LA	
43029	I	IWRP	LA	
43030	I	IWRP	PM	
43031	I	IWRP	PM	
43032	I	IWRP	PM	The Royal Regiment of Wales
43033	I	IWRP	PM	
43034	I	IWRP	PM	
43035	I	IWRP	PM	
43036	I	IWRP	PM	
43037	I	IWRP	PM	
43038	I	IECP	NL	National Railway Museum
				The First Ten Years 1975 – 1985
43039	I	IECP	NL	
43040	I	IWRP	PM	Granite City
43041	I	IWRP	PM	City of Discovery
43042	I	IWCP	LA	
43043	I	IMLP	NL	
43044	I	IMLP	NL	Borough of Kettering
43045	I	IMLP	NL	The Grammar School Doncaster AD 1350
43046	I	IMLP	NL	
43047	I	IMLP	NL	Rotherham Enterprise
43048	I	IMLP	NL	
43049	I	IMLP	NL	Neville Hill
43050	I	IMLP	NL	
43051	I	IMLP	NL	The Duke and Duchess of York
43052	I	IMLP	NL	City of Peterborough
43053	I	IMLP	NL	County of Humberside
43054	I	IMLP	NL	
43055	I	IMLP	NL	Sheffield Star
43056	I	IMLP	NL	University of Bradford
43057	I	IMLP	NL	Bounds Green
43058	I	IMLP	NL	
43059	I	IMLP	NL	
43060	I	IMLP	NL	County of Leicestershire
43061	I	IMLP	NL	City of Lincoln
43062	I	ICCS	EC	
43063	I	ICCS	EC	
43064	I	IMLP	NL	City of York
43065 §	I	ICCS	EC	
43066	I	IMLP	NL	
43067 §	I	ICCS	EC	
43068 §	I	ICCS	EC	
43069	I	ICCS	EC	
43070	I	ICCS	EC	
43071	I	ICCS	EC	
43072	I	IMLP	NL	Derby Etches Park
43073	I	IMLP	NL	
43074	I	IMLP	NL	

43075	I	IMLP	NL	
43076	I	IMLP	NL	BBC East Midlands Today
43077	I	IMLP	NL	County of Nottingham
43078	I	ICCS	EC	Shildon County Durham
43079	I	ICCS	EC	
43080 §	I	ICCS	EC	
43081	I	IMLP	NL	
43082	I	IMLP	NL	
43083	I	IMLP	NL	
43084 §	I	ICCS	EC	County of Derbyshire
43085	I	IMLP	NL	City of Bradford
43086	I	ICCP	NL	
43087	I	ICCP	NL	
43088	I	ICCP	NL	
43089	I	ICCP	NL	
43090	I	ICCS	EC	
43091	I	ICCS	EC	Edinburgh Military Tattoo
43092	I	ICCS	EC	Highland Chieftain
43093	I	ICCS	EC	York Festival '88
43094	I	ICCS	EC	
43095	I	IECP	NL	
43096	I	IECP	NL	The Queens Own Hussars
43097	I	ICCS	EC	
43098	I	ICCS	EC	
43099	I	ICCS	EC	
43100	I	ICCS	EC	Craigentinny
43101	I	ICCP	NL	
43102	I	ICCP	NL	
43103	I	ICCP	NL	John Wesley
43104	I	IECP	NL	County of Cleveland
43105	I	IECP	NL	Hartlepool
43106	I	IECP	NL	Songs of Praise
43107	I	IECP	NL	
43108	I	IECP	NL	
43109	I	IECP	NL	Yorkshire Evening Press
43110	I	IECP	NL	Darlington
43111	I	IECP	NL	
43112	I	IECP	NL	
43113	I	IECP	NL	City of Newcastle-upon-Tyne
43114	I	IECP	NL	National Garden Festival Gateshead 1990
43115	I	IECP	NL	Yorkshire Cricket Academy
43116	I	IECP	NL	City of Kingston Upon Hull
43117	I	IECP	NL	
43118	I	IECP	NL	Charles Wesley
43119	I	IECP	NL	
43120	I	IECP	NL	
43121	I	ICCP	NL	West Yorkshire Metropolitan County
43122	I	ICCP	NL	South Yorkshire Metropolitan County
43123 §	I	ICCS	EC	
43124	I	IWRP	PM	
43125	I	IWRP	PM	
43126	I	IWRP	PM	City of Bristol

43127		I	IWRP	PM	
43127		I	IWRP	PM	
43128		I	IWRP	PM	
43129		I	IWRP	PM	
43130		I	IWRP	PM	Sulis Minerva
43131		I	IWRP	PM	
43132		I	IWRP	PM	Worshipful Company of Carmen
43133		I	IWRP	PM	
43134		I	IWRP	PM	County of Somerset
43135		I	IWRP	PM	
43136		I	IWRP	PM	
43137		I	IWRP	PM	
43138		I	IWRP	PM	
43139		I	IWRP	PM	
43140		I	IWRP	OO	
43141		I	IWCP	OO	
43142		I	IWCP	OO	
43143		I	IWRP	LA	
43144		I	IWCP	OO	
43145		I	IWCP	OO	
43146		I	IWCP	OO	
43147		I	IWCP	OO	The Red Cross
43148		I	IWRP	OO	
43149		I	IWRP	PM	B.B.C. Wales Today
43150		I	IWRP	PM	Bristol Evening Post
43151		I	IWRP	PM	
43152		I	IWRP	PM	St. Peters School York AD 627
43153		I	ICCP	LA	University of Durham
43154		I	ICCP	LA	
43155		I	ICCP	LA	B.B.C. Look North
43156		I	ICCP	LA	
43157		I	ICCP	LA	Yorkshire Evening Post
43158		I	ICCP	LA	
43159		I	ICCP	LA	
43160		I	ICCP	LA	Storm Force
43161		I	ICCP	LA	Reading Evening Post
43162		I	ICCP	LA	Borough of Stevenage
43163		I	IWRP	LA	
43164		I	IWRP	LA	
43165		I	IWRP	LA	
43166		I	IWRP	LA	
43167	*	I	IWRP	PM	
43168	*	I	IWRP	PM	
43169	*	I	IWRP	PM	The National Trust
43170	*	I	ICCP	LA	
43171		I	IWRP	LA	
43172		I	IWRP	LA	
43173		I	IWRP	LA	
43174		I	IWRP	LA	
43175		I	IWRP	LA	
43176		I	IWRP	LA	
43177		I	IWRP	LA	
43178		I	IWRP	LA	

43179	I	IWRP	LA	Pride of Laira
43180	I	ICCP	NL	
43181	I	IWRP	LA	Devonport Royal Dockyard 1693-1993
43182	I	IWRP	LA	
43183	I	IWRP	LA	
43184	I	IWRP	LA	
43185	I	IWRP	LA	Great Western
43186	I	IWRP	LA	Sir Francis Drake
43187	I	IWRP	LA	
43188	I	IWRP	LA	City of Plymouth
43189	I	IWRP	LA	
43190	I	IWRP	LA	
43191	I	IWRP	LA	Seahawk
43192	I	IWRP	LA	City of Truro
43193	I	ICCP	LA	
43194	I	ICCP	LA	
43195	I	ICCP	LA	
43196	I	ICCP	LA	The Newspaper Society Founded 1836
43197	I	ICCP	LA	
43198	I	ICCP	NL	

CLASS 47   BRUSH TYPE 4   Co – Co

Built: 1963 – 67 by Brush Traction, Loughborough or BR Crewe Works.
Engine: Sulzer 12LDA28C of 1920 kW (2580 hp) at 750 rpm.
Main Generator: Brush TG160-60 Mk2, TG 160-60 Mk4 or TM172-50 Mk1.
Traction Motors: Brush TM64-68 Mk1 or Mk1A (axle hung).
Max. Tractive Effort: 267 kN (60000 lbf).
Cont. Tractive Effort: 133 kN (30000 lbf) at 26 mph.
Power At Rail: 1550 kW (2080 hp).   **Length over Buffers:** 19.38 m.
Brake Force: 61 t.   **Wheel Diameter:** 1143 mm.
Design Speed: 95 mph.   **Weight:** 120.5 – 125 t.
Max. Speed: various.   **RA:** 6 or 7.
Train Brakes: Air & vacuum.
Multiple Working: Not equipped (Blue Star Coupling Code*).
ETH Index (47/4 & 47/7): 66 (75§).
Communication Equipment: Cab to shore radio-telephone.

Non standard liveries:

47145 is BR blue with black cab surrounds and Railfreight general markings
47803 is grey, red and yellow.

a Vacuum brake isolated.

Formerly numbered 1100 – 11, 1500 – 1999 not in order. 47299 was previously
47216. 47300 was previously 47468.

Class 47/0. Built with train heating boiler. RA6. Max Speed 75 mph.

| | 47004 | G | IGJO | OC |
| | 47016 | F0 | IGJO | OC |

47019	FO	MDYX	DR (U)	
47033	a + FD	MDCT	TI	
47049	+ FD	MDDT	TI	
47050	a + FD	MDYX	TI (U)	
47051	a + FD	MDCT	TI	
47052	FD	MDAT	TI	
47053	a + FD	MDDT	TI	Cory Brothers 1842-1992
47060	a FD	MDYX	TI (U)	Halewood Silver Jubilee 1988
47063	FA	MDYX	TI (U)	
47079	FD	MDSR	TI (U)	
47085	+ F	MDDT	TI	REPTA 1893 – 1993
47095	+ FD	MDDT	TI	
47105		MDYX	DR (U)	
47108		IGJO	OC	
47114	a + FD	MDDT	TI	
47121		IGJO	OC	
47125	+ FE	MDDT	TI	
47142	FR	MDWT	TI	The Sapper
47144	a + FD	MDDT	TI	
47145	0	MDWT	TI	
47146	a	MDAT	TI	
47147	FD	MDAT	TI	
47150	+ FD	MDDT	TI	
47152	a + FD	MDDT	TI	
47156	a + FD	MDDT	TI	
47157	F	MDSR	TI (U)	
47186	a + FE	MDDT	TI	Catcliffe Demon
47187	FD	MDSR	TI (U)	
47188	a + FD	MDDT	TI	
47190	FP	MDYX	TI (U)	
47193	FP	MDWT	TI	
47194	+ FD	MDDT	TI	Carlisle Currock Quality Approved
47196	FD	MDSR	TI (S)	
47197	FP	MDSR	TI (S)	
47200	a + FD	MDDT	TI	
47201	+ FD	MDDT	TI	
47204	+ FD	MDDT	TI	
47205	+ FD	MDDT	TI	
47206	FD	MDAT	TI	
47207	FD	MDAT	TI	Bulmers of Hereford
47209	+ FD	MDDT	TI	Herbert Austin
47210	+ FD	MDSR	TI (U)	
47211	+ FD	MDDT	TI	
47212	+ FP	MDWT	TI	
47213	+ FD	MDDT	TI	Marchwood Military Port
47214	FD	MDYX	TI (U)	
47217	a + FE	MDDT	TI	
47218	+ FD	MDDT	TI	United Transport Europe
47219	a + FD	MDCT	TI	Arnold Kunzler
47221	+ FP	MDWT	TI	
47222	a + FD	MDCT	TI	
47223	+ FD	MDWT	*SL.*	

47224	+ **FP**	MDWT	TI	
47225	**FD**	MDYX	TI (U)	
47226	a + **FD**	MDDT	TI	
47228	+ **FD**	MDDT	TI	
47229	a + **FD**	MDDT	TI	
47231	**FD**	MDAT	TI	The Silcock Express
47234	+ **FE**	MDDT	TI	
47236	a + **FD**	MDCT	TI	
47237	a + **FD**	MDCT	TI	
47238	**FD**	MDAT	TI	Bescot Yard
47241	a + **FD**	MDDT	TI	
47245	+ **FD**	MDDT	TI	
47249	**FR**	MDWT	TI	
47256	**FD**	MDWT	TI	
47258	+ **FD**	MDDT	TI	
47270		MDWT	TI	
47276	+ **FP**	MDSR	TI (U)	
47277	**FP**	MDWT	TI	
+47278	**FP**	MDWT	TI *ISC*	
47279	**FD**	MDAT	TI	
47280	a + **FD**	MDCT	TI	Pedigree
47281	+ **FD**	MDDT	TI	
47283	**FD**	MDWT	TI	Johnnie Walker
47284	+ **FD**	MDDT	TI	
47285	a + **FD**	MDCT	TI	
47286	a + **F**	MDDT	TI	
47287	+ **FD**	MDDT	TI	
47288	**FD**	MDSR	TI (U)	
47289	a **FD**	MDSR	TI (U)	
47290	a + **FD**	MDDT	TI	
47291	a + **FD**	MDDT	TI	The Port of Felixstowe
47292	a + **FD**	MDDT	TI	
47293	a + **FD**	MDDT	TI	
47294	+ **FD**	MDWT	TI	
47295	+ **FP**	MDAT	TI	
47296	**FD**	MDAT	TI	
47297	a + **FD**	MDDT	TI	
47298	+ **FD**	MDDT	TI	Pegasus
47299	a + **FD**	MDDT	TI	

Class 47/3. Built without Train Heat. (except 47300). RA6. Max Speed 75 mph. All equipped with slow speed control.

47300	**C**	CEJC	BS	
47301	**FR**	MDAT	TI	
47302	a **FR**	MDAT	TI	
47303	a + **F**	MDDT	TI	
47304	a + **FD**	MDDT	TI	
47305	**FP**	MDAT	TI	
47306	a + **FD**	MDYX	TI (U)	
47307	a + **FE**	MDDT	TI	
47308	**F**	MDSR	TI (S)	
47309	+ **FD**	MDDT	TI	The Halewood Transmission

47310	a + **FD**	MDCT	TI	Henry Ford
47312	a + **FD**	MDDT	TI	
47313	a + **FD**	MDCT	TI	
47314	a + **FD**	MDDT	TI	Transmark
47315	**C**	IGJO	OC	Templecombe
47316	a + **FD**	MDCT	TI	
47317	**FD**	MDWT	TI	Willesden Yard
47319	+ **FP**	IEJW	IM	Norsk Hydro
47321	**F**	MDSR	TI (S)	
47322	**FR**	MDWT	TI	
47323	a + **FD**	MDCT	TI	
47325	**FO**	MDYX	TI (U)	
47326	a + **FD**	MDCT	TI	
47328	+ **FD**	MDDT	TI	
47329	**C**	IWJD	CD	
47330	a + **FD**	MDDT	TI	Amlwch Freighter/
				Trên Nwyddau Amlwch (Opp. Sides)
47331	**C**	IEJI	IM	
47332	**C**	CEJC	BS	
47333	**C**	CEJC	BS	Civil Link
47334	**C**	IWJD	CD	
47335	a + **FD**	MDDT	TI	
47337	**FO**	MDAT	TI	
47338	a + **FD**	MDDT	TI	Warrington Yard
47339	**FD**	MDAT	TI	
47340	**C**	IWJD	CD	
47341	**C**	CEJC	BS	
47344	a + **FD**	MDDT	TI	
47345	**FR**	MDWT	TI	
47346	**C**	RBJW	IM	
47347	a **FM**	MDAT	TI	
47348	**FO**	IMJC	BR	St. Christopher's Railway Home
47349	**FD**	MDYX	TI (U)	
47350	**FO**	MDWT	TI	
47351	a + **FD**	MDDT	TI	
47352	**C**	MDSR	TI (S)	
47353	**C**	CEJC	BS	
47354	a **FD**	MDAT	TI	
47355	+ **FD**	MDDT	TI	
47356	**FO**	CEJC	BS	
47357	**C**	CEJC	BS	The Permanent Way Institution
47358	**FO**	MDYX	SF (U)	
47359	**FD**	MDAT	TI	
47360	a + **FD**	MDDT	TI	
47361	a + **FD**	MDDT	TI	Wilton Endeavour
47362	a + **FD**	MDCT	TI	
47363	+ **F**	MDDT	TI	
47365	+ **FD**	MDDT	TI	ICI Diamond Jubilee
47366	**FO**	IGJO	OC	
47367	**FR**	MDAT	TI	
47368	**FP**	IMJC	BR	
47369	**FD**	MDWT	TI	

47370	**FO**	MDSR	TI (U)	
47371	**FO**	MDSR	TI (S)	
47372	**C**	CEJC	BS	
47375 a + **FD**		MDDT	TI	Tinsley Traction Depot
				Quality Approved
47376		MDSR	TI (U)	
47377 a	**FD**	MDAT	TI	
47378 a + **FD**		MDDT	TI	
47379 +	**FP**	MDWT	TI	

Class 47/4. Equipped with train heating. RA7. Max Speed 95 (75§) mph.

47462	**R**	IMJC	BR	Cambridge Traction & Rolling Stock
				Depot
47463		PXLH	CD	
47467	**BR**	PXLC	CD	
47471	**IO**	PXLH	CD	Norman Tunna G.C.
47473	**BR**	IWJD	CD	
47474	**R**	PXLC	CD	Sir Rowland Hill
47475	**RX**	PXLC	CD	Restive
47476	**R**	PXLC	CD	Night Mail
47478		IWJD	CD	
47481	**BR**	PXLH	CD	
47484	**G**	IGJO	OC	ISAMBARD KINGDOM BRUNEL
47489	**R**	PXLC	CD	Crewe Diesel Depot
47490 +	**RX**	PXLB	CD	Resonant
47491 +	**RX**	PXLB	CD	Resolve
47492	**IO**	PXLH	CD	
47500 +	**RX**	PXLB	CD	
47501	**R**	PXLH	CD	Craftsman
47503 +	**RX**	PXLB	CD	Heaton Traincare Depot
47513	**BR**	PXLH	CD	Severn
47517 +	**RX**	PXLB	CD	
47519	**BR**	PXLH	CD	
47520	**M**	IEJT	IM	Thunderbird
47521	**RX**	PXLC	CD	
47522	**R**	PXLH	CD	Doncaster Enterprise
47523	**M**	PXLC	CD	
47524	**M**	PXLC	CD	
47525	**IO**	IWJD	CD	
47526	**BR**	NXXB	SF	
47528	**M**	PXLC	CD	The Queen's Own Mercian
				Yeomanry
47530	**RX**	PXLC	CD	
47532	**RX**	PXLC	CD	
47535	**R**	PXLH	CD	University of Leicester
47536	**BR**	PXLH	CD	
47537 +	**RX**	PXLB	CD	
47541 +	**RX**	PXLB	CD	
47543	**R**	PXLC	CD	
47547	**N**	PXLH	CD	
47550	**M**	IEJW	IM	University of Dundee

47555 (47126)	**IO**	IMJC	BR	
47557 (47024)	**RX**	PXLC	CD	
47558 (47027)	**RX**	PXLC	CD	
47559 (47028) +	**RX**	PXLB	CD	
47562 (47672) +	**RX**	PXLB	CD	Restless
47564 (47038) +	**RX**	PXLB	CD	
47565 (47039)	**RX**	PXLC	CD	Responsive
47566 (47043)	**M**	PXLC	CD	
47567 (47044)	**RX**	PXLC	CD	Red Star
47568 (47045)	**RX**	PXLC	CD	Royal Logistic Corps Postal & Courier Services
47569 (47047) +	**RX**	PXLB	CD	The Gloucestershire Regiment
47572 (47168)	**R**	PXLC	CD	Ely Cathedral
47573 (47173) +	**RX**	PXLC	CD	
47574 (47174)	**R**	PXLC	CD	Benjamin Gimbert G.C.
47575 (47175)	**R**	PXLC	CD	City of Hereford
47576 (47176)	**RX**	PXLC	CD	
47578 (47181) +	**RX**	PXLB	CD	
47579 (47183)	**N**	NXXB	SF	James Nightall G.C.
47580 (47167) +	**RX**	PXLB	CD	
47581 (47169) +	**RX**	PXLB	CD	
47582 (47170)	**R**	PXLC	CD	County of Norfolk
47583 (47172)	**RX**	PXLC	CD	
47584 (47180)	**RX**	PXLC	CD	County of Suffolk
47587 (47263)	**RX**	PXLC	CD	
47588 (47178)	**RX**	PXLC	CD	Resurgent
47592 (47171) +	**RX**	PXLB	CD	Resourceful
47594 (47035)	**RX**	PXLC	CD	Resourceful
47596 (47255)	**RX**	PXLC	CD	
47597 (47026) +	**RX**	PXLB	CD	Resilient
47598 (47182)	**RX**	PXLC	CD	
47599 (47177)	**RX**	PXLC	CD	
47600 (47250)	**RX**	PXLC	CD	
47603 (47267) +	**RX**	PXLB	CD	
47605 (47160)	**RX**	PXLC	CD	
47612 (47838) +	**RX**	PXLB	CD	
47615 (47252)	**RX**	PXLC	CD	
47618 (47836) +	**RX**	PXLB	CD	
47624 (47087)	**M**	PXLC	CD	
47625 (47076)	**RX**	PXLC	CD	Resplendent
47626 (47082)	**M**	PXLC	CD	ATLAS
47627 (47273)	**RX**	PXLC	CD	
47628 (47078)	**RX**	PXLC	CD	
47630 (47041) +	**RX**	PXLB	CD	Resounding
47631 (47059) +	**RX**	PXLB	CD	Ressaldar
47634 (47158)	**R**	PXLC	CD	Holbeck
47635 (47029)	**R**	PXLC	CD	
47636 (47243) +	**RX**	PXLB	CD	Restored
47040 (17244)	**R**	PXLC	CD	University of Strathclyde
47641 (47086) +	**RX**	PXLB	CD	
47642 (47040) +	**RX**	PXLB	CD	Resolute

47644	(47246)	+	**RX**	PXLB	CD	
47653	(47808)	+	**RX**	PXLB	CD	
47671	(47616)	§	**BR**	IEJT	IM	
47673	(47593)	§	**IO**	IEJT	IM	Galloway Princess
47674	(47604)	§	**BR**	IMJC	BR	Women's Royal Voluntary Service
47675	(47595)	§	**M**	IEJT	IM	Confederation of British Industry
47676	(47586)	§	**I**	IEJI	IM	Northamptonshire
47677	(47617)	§	**I**	IMJC	BR	University of Stirling

Class 47/7. Fitted with an older form of TDM. RA6. Max speed 100 mph.

47701	(47493)	+	**RX**	PXLB	CD	
47702	(47504)	+	**N**	NXXB	SF	Saint Cuthbert
47703	(47514)	+	**R**	PXLB	CD	The Queen Mother
47704	(47495)	+	**RX**	PXLB	CD	
47705	(47554)	+	**RX**	PXLB	CD	
47706	(47494)	+	**PS**	PXLB	CD	
47707	(47506)	+	**RX**	PXLB	CD	Holyrood
47708	(47516)	+	**N**	PXLD	CD	
47709	(47499)	+	**RX**	PXLB	CD	
47710	(47496)	+	**N**	PXLB	CD	
47711	(47498)	+	**N**	NXXB	SF	County of Hertfordshire
47712	(47505)	+	**R**	PXLB	CD	Lady Diana Spencer
47714	(47511)	+	**N**	PXLB	CD	
47715	(47502)	+	**N**	PXLB	CD	Haymarket
47716	(47507)	+	**N**	PXLB	CD	Duke of Edinburgh's Award
47717	(47497)	+	**R**	PXLB	CD	

Class 47/4 continued.

47721	(	)
47722	(	)
47723	(	)
47724	(	)
47725	(	)
47726	(	)
47727	(	)
47728	(	)
47729	(	)
47730	(	)
47731	(	)
47732	(	)
47733	(	)
47734	(	)
47735	(	)
47736	(	)
47737	(	)
47738	(	)
47739	(	)
47740	(	)
47741	(	)
47742	(	)

47743 (	)				
47744 (	)				
47745 (	)				
47746 (	)				
47747 (	)				
47748 (	)				
47749 (	)				
47750 (	)				
47751 (	)				
47752 (	)				
47753 (	)				
47754 (	)				
47755 (	)				
47756 (	)				
47757 (	)				
47758 (	)				
47759 (	)				
47760 (	)				
47761 (	)				
47762 (	)				
47763 (	)				
47764 (	)				
47765 (	)				
47766 (	)				
47767 (	)				
47768 (	)				
47769 (	)				
47770 (	)				
47771 (	)				
47772 (	)				
47773 (	)				
47774 (47551)	+	**RX**	PXLB	CD	Poste Restante
47775 (47531)	+	**RX**	PXLB	CD	Respite
47776 (	)				
47777 (	)				
47778 (47606)	+	**RX**	PXLB	CD	Irresistible
47779 (	)				
47780 (	)				
47781 (	)				
47802 (47552)	+	I	IMJC	BR	
47803 (47553)	+	0	IMJC	BR	
47804 (47591)	+	I	IMJC	BR	Kettering
47805 (47650)	+	I	ILRA	BR	Bristol Bath Road
47806 (47651)	+	I	ILRA	BR	
47807 (47652)	+	I	ILRA	BR	
47809 (47654)	+	I	IBRA	BR	Finsbury Park
47810 (47655)	+	I	ILRA	BR	
47811 (47656)	+	I	ILRA	BR	
47812 (47657)	+	I	ILRA	BR	
47813 (47658)	+	I	ILRA	BR	
47814 (47659)	+	I	ILRA	BR	
47815 (47660)	+	I	ILRA	BR	

47816	(47661)	+	I	ILRA	BR	
47817	(47662)	+	I	ILRA	BR	
47818	(47663)	+	I	ILRA	BR	
47819	(47664)	+	I	IBRA	BR	
47820	(47665)	+	I	IBRA	BR	
47821	(47607)	+	I	IBRA	BR	Royal Worcester
47822	(47571)	+	I	ILRA	BR	
47823	(47610)	+	I	IBRA	BR	SS Great Britain
47824	(47602)	+	M	PXLB	CD	
47825	(47590)	+	M	ILRA	BR	Thomas Telford
47826	(47637)	+	I	ILRA	BR	
47827	(47589)	+	I	ILRA	BR	
47828	(47629)	+	I	ILRA	BR	
47829	(47619)	+	I	ILRA	BR	
47830	(47649)	+	I	ILRA	BR	
47831	(47563)	+	M	ILRA	BR	Bolton Wanderer
47832	(47560)	+	M	ILRA	BR	Tamar
47833	(47608)	+	G	IBRA	BR	Captain Peter Manisty RN
47834	(47609)	+	I	IBRA	BR	FIRE FLY
47835	(47620)	+	I	IBRA	BR	Windsor Castle
47839	(47621)	+	I	ILRA	BR	
47840	(47613)	+	I	IBRB	BR (S)	NORTH STAR
47841	(47622)	+	I	ILRA	BR	The Institution of Mechanical Engineers
47843	(47623)	+	I	IBRB	BR (S)	
47844	(47556)	+	I	ILRA	BR	Derby & Derbyshire Chamber of Commerce & Industry
47845	(47638)	+	I	ILRA	BR	County of Kent
47846	(47647)	+	I	ILRA	BR	THOR
47847	(47577)	+	I	ILRA	BR	
47848	(47632)	+	I	ILRA	BR	
47849	(47570)	+	M	ILRA	BR	
47850	(47648)	+	I	ILRA	BR	
47851	(47639)	+	I	IBRB	BR (S)	
47853	(47614)	+	M	ILRA	BR	
47971	(97480)	*	BR	CDJC	BS	Robin Hood
47972	(97545)		CS	CDJC	BS	The Royal Army Ordnance Corps
47973	(97561)		M	CDJC	BS	Derby Evening Telegraph
47975	(47540)	*	C	CDJC	BS	The Institution of Civil Engineers
47976	(47546)	*	C	CDJC	BS	Aviemore Centre

Class 47/3 continued. 100 mph loco for research purposes.

47981	(47364)		C	CDJC	BS	

CLASS 50   ENGLISH ELECTRIC TYPE 4   Co-Co

Built: 1967 – 68 by English Electric Co. at Vulcan Foundry, Newton le Willows
Engine: English Electric 16CVST of 2010 kW (2700 hp) at 850 rpm.
Main Generator: English Electric 840/4B.
Traction Motors: English Electric 538/5A.
Max. Tractive Effort: 216 kN (48500 lbf).

Cont. Tractive Effort: 147 kN (33000 lbf) at 23.5 mph.
Power At Rail: 1540 kW (2070 hp). **Length over Buffers:** 20.88 m.
Brake Force: 59 t. **Wheel Diameter:** 1092 mm.
Design Speed: 105 mph. **Weight:** 117 t.
Max. Speed: 100 mph. **RA:** 6.
Train Heating: Electric. **ETH Index:** 61.
Train Brakes: Air & vacuum.
Multiple Working: Orange Square coupling code. (Within class only).
Communication Equipment: Cab to shore radio-telephone.
All equipped with slow speed control.

Formerly numbered 407, 433, 400. 50050 carries D 400.

50007	G	NWXA	LA	SIR EDWARD ELGAR
50033	N	NWXA	LA	Glorious
50050		NWXA	LA	

CLASS 56   BRUSH TYPE 5   Co–Co

Built: 1976 – 84 by Electroputere at Craiova, Romania (as sub contractors for Brush) or BREL at Doncaster or Crewe Works.
Engine: Ruston Paxman 16RK3CT of 2460 kW (3250 hp) at 900 rpm.
Main Alternator: Brush BA1101A.
Traction Motors: Brush TM73-62.
Max. Tractive Effort: 275 kN (61800 lbf).
Cont. Tractive Effort: 240 kN (53950 lbf) at 16.8 mph.
Power At Rail: 1790 kW (2400 hp). **Length over Buffers:** 19.36 m.
Brake Force: 60 t. **Wheel Diameter:** 1143 mm.
Design Speed: 80 mph. **Weight:** 125 t.
Max. Speed: 80 mph. **RA:** 7.
Train Brakes: Air.
Multiple Working: Red Diamond coupling code.
Communication Equipment: Cab to shore radio-telephone.
All equipped with slow speed control.

§ Derated to 1790 kW (2400 hp).
* Derated to 2060 kW (2800 hp).

56001	FA	FQXB	SL	Whatley
56003	F	FCCI	IM	
56004		FCBN	TO	
56005	FC	FCDN	TO	
56006	FC	FCDN	TO	
56007	FC	FCBN	TO	
56008		FCXX	TO (U)	
56009	FC	FCBN	TO	
56010		FCBN	TO	
56011	FR	FCDN	TO	
56012	FC	FCXX	TO (U)	
56013	FC	FCXX	TO (U)	
56014	FC	FCXX	TO	
56016	FC	FCXX	TO (U)	
56018	FC	FCNN	TO	
56019	FR	FCNN	TO	

56020		FCXX	TO (U)	
56021	FC	FCDN	TO	
56022		FCXX	TO (U)	
56023	FC	FCXX	TO (U)	
56024	FO	FCXX	TO (U)	
56025	FC	FCPA	TO	
56026		FCXX	TO (U)	
56027	FC	FCXX	TO (U)	
56029	FC	FCNN	TO	
56031	C	NKJM	SL	Merehead
56032	FM	FMCK	CF	Sir De Morgannwg/ County of South Glamorgan
56033	FA	FASB	SL	
56034	FA	FMTY	TE	Castell Ogwr/Ogmore Castle
56035	FA	FASB	SL	
56036	C	NKJM	SL	
56037	FA	FASB	SL	Richard Trevithick
56038	FM	FMCK	CF	Western Mail
56039	FM	FMTY	TE	
56040	FM	FMCK	CF	Oystermouth
56041	FA	FASB	SL	
56043	FA	FMTY	TE	
56044	FM	FMCK	CF	Cardiff Canton Quality Assured
56045	FA	FMTY	TE	
56046	C	NKJM	SL	
56047	C	NKJM	SL	
56048	C	NKJM	SL	
56049	C	NKJM	SL	
56050	FA	FMTY	TE	
56051	FA	FASB	SL	Isle of Grain
56052	FM	FMXX	CF	
56053	FA	FMCK	CF	Sir Morgannwg Ganol/ County of Mid Glamorgan
56054	FM	FMCK	CF	British Steel Llanwern
56055	FA	FASB	SL	
56056	FA	FASB	SL	
56057	FA	FASB	SL	
56058	FA	FASB	SL	
56059	FA	FASB	SL	
56060	FM	FMCK	CF	The Cardiff Rod Mill
56061	FM	FMTY	TE	
56062	FA	FCEN	TO	Mountsorrel
56063	FA	FMTY	TE	Bardon Hill
56064	FM	FMCK	CF	
56065	FA	FASB	SL	
56066	FC	FCXX	TO	
56067	FC	FCDN	TO	
56068	FC	FCDN	TO	
56069	§ FM	FMTY	TE	Thornaby TMD
56070	FA	FASB	SL	
56071	FC	FCDN	TO	
56072	F	FCDN	TO	

THE PLATFORM 5
TRANSPORT BOOK CLUB

The Platform 5 Transport Book Club is a service for transport enthusiasts which enables members to order new Platform 5 titles before publication, at discounts of between 15% and 25% of the normal retail price. Unlike other book clubs, a small annual subscription fee of £1.50 is charged to cover the production and postage of a quarterly newsletter, but there is no obligation whatsoever to buy any books at any time. We like customers to buy our books because of their quality, not because of any obligation to a book club.

To illustrate the sort of discounts available, the following offer applied to Exeter – Newton Abbot – A Railway History before publication in November 1993:

```
Cover Price:. . . . . . . . . . . . . . . . . . . . . . . . . . . .£25.00
Less Book Club Member Discount:. . . . . . . . . . . .£5.00

SUB-TOTAL . . . . . . . . . . . . . . . . . . . . . . . . . . . . £20.00

Plus 10% Contribution to Postage & Packing . . . . .£2.00

BOOK CLUB MEMBER PRICE. . . . . . . . . . . . . . .£22.00
```

Even after our 10% postage and packing charge, this still represents a saving of £3.00 on the cover price.

Discounted prices will of course, only be available to Platform 5 Transport Book Club members, and will only be applicable if subscriptions are received by a certain date prior to publication, to be advised.

In addition, a discount on selected existing Platform 5 titles will occasionally be offered to Book Club members.

There is no limit to the number of copies of each book that may be ordered through the Platform 5 Transport Book Club.

Books ordered through the Platform 5 Transport Book Club will be despatched as soon as possible after publication.

Customers should be aware that although most new Platform 5 titles will be available via this service, some low-priced titles will be excluded and in particular, the British Rail Pocket Books will not be available at a discounted price. These books may still be ordered through the club, at the normal retail price.

THE PLATFORM 5 TRANSPORT BOOK CLUB
MEMBERSHIP APPLICATION FORM

To enrol for one year's membership in the PLATFORM 5 TRANSPORT BOOK CLUB, please complete this form (or a photocopy) and send it with your cheque/postal order for £1.50 made payable to 'Platform 5 Publishing Limited' to:

The Platform 5 Transport Book Club, Wyvern House, Sark Road, SHEFFIELD, S2 4HG.

BLOCK CAPITALS PLEASE

Name: . . . .

Address: . . . .

. . . .

Post Code: . . . .

Telephone: . . . .

Please accept my application and enrol me as a member of the Platform 5 Transport Book Club.

As a member I will receive four issues of the club newsletter, each containing a number of new books at prices of at least 15% less than the published cover price (exclusive of postage and packing).

I understand I am not obliged to buy any of the books offered, and there is no limit to the number of copies of each book that may be ordered.

I enclose my cheque/postal order for £1.50 payable to Platform 5 Publishing Limited.

Signed: . . . .

Date: . . . .

Office Use Only: . . . .

PLATFORM 5 PUBLISHING LTD.
MAIL ORDER LIST

NEW TITLES	Price
BR Pocket Book No. 1: Locomotives	1.85
BR Pocket Book No. 2: Coaching Stock	1.85
BR Pocket Book No. 3: DMUs & Channel Tunnel Stock	1.85
BR Pocket Book No. 4: EMUs	1.85
British Railways Locomotives & Coaching Stock 1994 **MAR 94**	7.50
Light Rail Review 5	7.50
Preserved Locomotives of British Railways 8th edition	6.95
Exeter – Newton Abbot – A Railway History	25.00
Manx Electric	8.95
Steam Alive (Friends of the NRM)	2.95
Departmental Coaching Stock 5th edition (SCTP)	6.95
Buses in Britain (Capital)	19.95
Going Green (Capital)	5.95
TGV Handbook (Capital)	7.95

Modern British Railway Titles

On-Track Plant on British Railways 4th Edition	5.50
The Fifty 50s in Colour	5.95
Blood, Sweat and Fifties (Class 50 Society)	2.95
British Rail Internal Users (SCTP)	7.95
British Rail Wagon Fleet – Air Braked Stock (SCTP)	6.95
British Rail Wagon Fleet Volume 5 (SCTP)	5.25
RIV Wagon Fleet (SCTP)	5.95
British Rail Passenger Trains (Capital)	7.95
This is London Transport (Capital)	4.95
Miles & Chains Volume 2 – London Midland (Milepost)	1.40
Miles & Chains Vol. 3 - Scottish, Vol. 5 - Southern (Milepost)	each 1.00
Class Fifty Factfile (Class 50 Society)	2.95

Overseas Railways

German Railways Locomotives & MUs 3rd edition	12.50
Swiss Railways/Chemins de Fer Suisses	9.95
French Railways/Chemins de Fer Francais 2nd edition	9.95
ÖBB/Austrian Federal Railways 2nd edition	6.95
Benelux Locomotives & Coaching Stock 2nd edition	6.95
A Guide to Portuguese Railways (Fearless)	4.95
Paris Metro Handbook (Capital)	7.95
World Metro Systems (Capital)	6.95

Historical Railway Titles

6203 'Princess Margaret Rose'	19.95
Midland Railway Portrait	12.95
Steam Days on BR 1 – The Midland Line in Sheffield	4.95
Rails along the Sea Wall (Dawlish – Teignmouth Pictorial)	4.95
The Rolling Rivers	6.95
British Baltic Tanks	6.95
The Railways of Winchester	6.95
Register of Closed Railways 1948–91 (Milepost)	5.95
LNWR Branch Lines of West Leics & East Warwicks (Milepost)	7.95
British Railways Mark 1 Coaches (Atlantic)	19.95

Mallard – The Record Breaker (Friends of the NRM)............................4.95
Rails Through The Clay (Capital)..25.00
The 1938 Tube Stock (Capital)..9.95
Metropolitan Steam Locomotives (Capital).....................................9.95
The First Tube (Capital)...4.95
Register of Closed Railways 1948 – 91 (Milepost).............................5.95

Political

The Battle for the Settle & Carlisle...6.95

Rambling

Rambles by Rail 1 – The Hope Valley Line...................................1.95
Rambles by Rail 2 – Liskeard-Looe...1.95
Rambles by Rail 4 – The New Forest..1.95

Light Rail Transit, Trams & Buses

Light Rail Review 1 (Reprint)..6.95
Light Rail Review 2/3/4..each 7.50
UK Light Rail Systems No.1: Manchester Metrolink.............................8.50
Blackpool & Fleetwood By Tram..7.50
London Bus Handbook Part 1 (Capital)...8.95
Bus Handbook 9: Wales (Capital)..7.95
Bus Handbook 10: Scotland (Capital)..8.95
Routemaster Handbook (Capital)...7.95
Bus Review 8 (Bus Enthusiast)..4.95
Edinburgh's Trams & Buses (Bus Enthusiast)...................................4.95

Maps and Track Diagrams (Quail Map Company)

British Rail Track Diagrams 1 – Scotland & Isle of Man.....................5.00
British Rail Track Diagrams 3 – Western....................................5.00
British Rail Track Diagrams 4 – London Midland.............................6.95
London Railway Map...5.95
London Transport Track Map...1.30
China Railway Atlas..3.50
Czech Republic & Slovakia Railway Map..1.70
Greece Railway Map...1.00
Poland Railway Map...2.00

PVC Book Covers

A6 Pocket Book Covers in Blue, Red, Green or Grey............................0.80
Locomotives & CS Covers in Blue, Red, Green or Grey..........................1.00
A5 Book Covers in Blue, Red, Green or Grey...................................1.40

Back Numbers

Locomotives & Coaching Stock 1985..2.95
Locomotives & Coaching Stock 1986..3.30
Locomotives & Coaching Stock 1987..3.30
Locomotives & Coaching Stock 1988..3.95
Locomotives & Coaching Stock 1989..4.95
Locomotives & Coaching Stock 1990..5.95
Locomotives & Coaching Stock 1991..6.60
British Railways Locomotives & Coaching Stock 1992...........................7.00
British Railways Locomotives & Coaching Stock 1993...........................7.25

Postage: 10% (UK), 20% (Europe), 30% (Rest of World). Minimum 30p.

All these publications are available from shops, bookstalls or direct from: Mail Order Department, Platform 5 Publishing Ltd., Wyvern House, Sark Road, SHEFFIELD, S2 4HG, ENGLAND. For a full list of titles available by mail order, please send SAE to the above address.

56073	**FM**	FMCK	CF	Tremorfa Steelworks
56074	**FC**	FCDN	TO	Kellingley Colliery
56075	**FC**	FCDN	TO	West Yorkshire Enterprise
56076	**FM**	FMCK	CF	British Steel Trostre
56077	§ **FC**	FCDN	TO	Thorpe Marsh Power Station
56078	**FC**	FCDN	TO	
56079	**FC**	FCDN	TO	
56080	**FC**	FCDN	TO	Selby Coalfield
56081	**FC**	FCEN	TO	
56082	**FC**	FCDN	TO	
56083	* **FC**	FCDN	TO	
56084	* **FC**	FPGI	IM	
56085	**FC**	FCCI	IM	
56086	* **FC**	FCDN	TO	
56087	**FM**	FMTY	TE	
56088	**FC**	FCCI	IM	
56089	**FC**	FCCI	IM	Ferrybridge C Power Station
56090	**FC**	FCCI	IM	
56091	**FC**	FCDN	TO	Castle Donington Power Station
56092	**F**	FCNN	TO	
56093	**FC**	FCDN	TO	The Institution of Mining Engineers
56094	**FC**	FPGI	IM	Eggborough Power Station
56095	**FC**	FCDN	TO	Harworth Colliery
56096	**FC**	FCDN	TO	
56097	**FM**	FMTY	TE	
56098	**FC**	FCDN	TO	
56099	**FC**	FCNN	TO	Fiddlers Ferry Power Station
56100	**FC**	FCDN	TO	
56101	**FC**	FCDN	TO	Mutual Improvement
56102	**FC**	FCDN	TO	Scunthorpe Steel Centenary
56103	**FA**	FASB	SL	
56104	**FC**	FCPA	TO	
56105	**FA**	FASB	SL	
56106	**FC**	FPGI	IM	
56107	§ **FC**	FCEN	TO	
56108	**FR**	FCEN	TO	
56109	**FC**	FCEN	TO	
56110	**FA**	FCEN	TO	Croft
56111	**FC**	FCEN	TO	
56112	**FC**	FCEN	TO	
56113	**FC**	FCKK	CF	
56114	**FC**	FCKK	CF	Maltby Colliery
56115	**FC**	FCKK	CF	
56116	**FC**	FMTY	TE	
56117	**FC**	FCEN	TO	Wilton-Coalpower
56118	**FC**	FCEN	TO	
56119	**FC**	FCKK	CF	
56120	**FC**	FCEN	TO	
56121	**FC**	FCPA	TO	
56123	**FC**	FCPA	TO	Drax Power Station
56124	**FC**	FCPA	TO	
56125	**FC**	FCPA	TO	

56126	FC	FPGI	IM	
56127	FC	FCPA	TO	
56128	FC	FCPA	TO	West Burton Power Station
56129	FC	FCPA	TO	
56130	FC	FCEN	TO	Wardley Opencast
56131	FC	FCEN	TO	Ellington Colliery
56132	FC	FCNN	TO	
56133	FC	FCEN	TO	Crewe Locomotive Works
56134	FC	FCEN	TO	Blyth Power
56135	F	FCEN	TO	Port of Tyne Authority

CLASS 58   BREL TYPE 5   Co – Co

Built: 1983 – 87 by BREL at Doncaster Works.
Engine: Ruston Paxman RK3ACT of 2460 kW (3300 hp) at 1000 rpm.
Main Alternator: Brush BA1101B.
Traction Motors: Brush TM73-62.
Max. Tractive Effort: 275 kN (61800 lbf).
Cont. Tractive Effort: 240 kN (53950 lbf) at 17.4 mph.
Power At Rail: 1780 kW (2387 hp).　**Length over Buffers:** 19.13 m.
Brake Force: 62 t.　　　　　　　　 **Wheel Diameter:** 1120 mm.
Design Speed: 80 mph.　　　　　　 **Weight:** 130 t.
Max. Speed: 80 mph.　　　　　　　 **RA:** 7.
Train Brakes: Air.
Multiple Working: Red Diamond coupling code.
Communication Equipment: Cab to shore radio-telephone.
All equipped with slow speed control.

58001	FC	FCBN	TO	
58002	FC	FCBN	TO	Daw Mill Colliery
58003	FC	FCBN	TO	Markham Colliery
58004	FC	FCBN	TO	
58005	FC	FCBN	TO	
58006	FC	FCBN	TO	
58007	FC	FCBN	TO	Drakelow Power Station
58008	FC	FCBN	TO	
58009	FC	FCBN	TO	
58010	FC	FCBN	TO	
58011	FC	FCBN	TO	Worksop Depot
58012	FC	FCBN	TO	
58013	FC	FCBN	TO	
58014	FC	FCBN	TO	Didcot Power Station
58015	FC	FCBN	TO	
58016	FC	FCBN	TO	
58017	FC	FCBN	TO	
58018	FC	FCBN	TO	High Marnham Power Station
58019	FC	FCBN	TO	Shirebrook Colliery
58020	FC	FCBN	TO	Doncaster Works
58021	FC	FCBN	TO	
58022	FC	FCBN	TO	
58023	FC	FCBN	TO	
58024	FC	FCBN	TO	

58025	FC	FCBN	TO	
58026	FC	FCBN	TO	
58027	FC	FCBN	TO	
58028	FC	FCBN	TO	
58029	FC	FCBN	TO	
58030	FC	FCBN	TO	
58031	FC	FCBN	TO	
58032	FC	FCBN	TO	
58033	FC	FCBN	TO	
58034	FC	FCBN	TO	Bassetlaw
58035	FC	FCBN	TO	
58036	FC	FCBN	TO	
58037	FC	FCBN	TO	
58038	FC	FCBN	TO	
58039	FC	FCDN	TO	Rugeley Power Station
58040	FC	FCDN	TO	Cottam Power Station
58041	FC	FCBN	TO	Ratcliffe Power Station
58042	FC	FCDN	TO	Ironbridge Power Station
58043	FC	FCDN	TO	Knottingley
58044	FC	FCBN	TO	Oxcroft Opencast
58045	FC	FCDN	TO	
58046	FC	FCDN	TO	Thoresby Colliery
58047	FC	FCDN	TO	Manton Colliery
58048	FC	FCDN	TO	Coventry Colliery
58049	FC	FCDN	TO	Littleton Colliery
58050	FC	FCBN	TO	Toton Traction Depot

CLASS 59   GENERAL MOTORS TYPE 5   Co – Co

Built: 1985 (59001 – 4), 1989 (59005) by General Motors, La Grange, Illinois, U.S.A. or 1990 (59101 – 4) by General Motors, London, Ontario, Canada.
Engine: General Motors 645E3C two stroke of 2460 kW (3300 hp) at 900 rpm.
Main Alternator: General Motors AR11 MLD-D14A.
Traction Motors: General Motors D77B.
Max. Tractive Effort: 506 kN (113 550 lbf).
Cont. Tractive Effort: 291 kN (65 300 lbf) at 14.3 mph.
Power At Rail: 1889 kW (2533 hp).   **Length over Buffers:** 21.35 m.
Brake Force: 69 t.   **Wheel Diameter:** 1067 mm.
Design Speed: 60 mph.   **Weight:** 121 t.
Max. Speed: 60 mph.   **RA:** 7.

Class 59/0. Owned by Foster-Yeoman Ltd. Blue/silver/blue livery with white lettering and cast numberplates.

59001	0	XYPO	FY	YEOMAN ENDEAVOUR
59002	0	XYPO	FY	YEOMAN ENTERPRISE
59003	0	XYPO	FY	YEOMAN HIGHLANDER
59004	0	XYPO	FY	YEOMAN CHALLENGER
59005	0	XYPO	FY	KENNETH J. PAINTER

Class 59/1. Owned by ARC Limited. Yellow/grey with grey lettering.

59101	0	XYPA	AR	Village of Whatley
59102	0	XYPA	AR	Village of Chantry

| 59103 | **0** | XYPA | AR | Village of Mells |
| 59104 | **0** | XYPA | AR | Village of Great Elm |

Class 59/2. Under construction for National Power.

| 59201 | **0** |

CLASS 60   BRUSH TYPE 5   Co–Co

Built: 1989 onwards by Brush Traction.
Engine: Mirrlees MB275T of 2310 kW (3100 hp) at 1000 rpm.
Main Alternator: Brush .
Traction Motors: Brush separately excited.
Max. Tractive Effort: 500 kN (106500 lbf).
Cont. Tractive Effort: 336 kN (71570 lbf) at 17.4 mph.
Power At Rail: 1800 kW (2415 hp). **Length over Buffers:** 21.34 m.
Brake Force: 74 t.   **Wheel Diameter:** 1118 mm.
Design Speed: 62 mph.   **Weight:** 129 t.
Max. Speed: 60 mph.   **RA:** 7.
Multiple Working: Within class.
Communication Equipment: Cab to shore radio-telephone.
All equipped with slow speed control.

— 60001	**FA**	FASB	SL	Steadfast
60002	**FP**	FPDI	IM	Capability Brown
60003	**FP**	FPDI	IM	Christopher Wren
60004	**FC**	FCJN	TO	Lochnagar
60005	**FA**	FABI	IM	Skiddaw
⊝ 60006	**FA**	FAXN	TO	Great Gable
60007	**FP**	FPTY	TE	Robert Adam
60008	**FM**	FPDI	IM	Moel Fammau
⊘ 60009	**FA**	FAXN	TO	Carnedd Dafydd
⊙ 60010	**FA**	FAXN	TO	Pumlumon Plynlimon
⊙ 60011	**FA**	FAXN	TO	Cader Idris
⊙ 60012	**FA**	FAXN	TO	Glyder Fawr
60013	**FP**	FPDI	IM	Robert Boyle
60014	**FP**	FPDI	IM	Alexander Fleming
60015	**FA**	FABI	IM	Bow Fell
60016	**FA**	FABI	IM	Langdale Pikes
⊙ 60017	**FA**	FASB	SL	Arenig Fawr
— 60018	**FA**	FASB	SL	Moel Siabod
— 60019	**FA**	FASB	SL	Wild Boar Fell
60020	**FM**	FMMY	TE	Great Whernside
60021	**FM**	FPDI	IM	Pen-y-Ghent
60022	**FM**	FMMY	TE	Ingleborough
60023	**FM**	FMMY	TE	The Cheviot
60024	**FP**	FPDI	IM	Elizabeth Fry
60025	**FP**	FPDI	IM	Joseph Lister
60026	**FP**	FPDI	IM	William Caxton
60027	**FP**	FPDI	IM	Joseph Banks
60028	**FP**	FPDI	IM	John Flamsteed
60029	**FM**	FMEK	CF	Ben Nevis

60030	FM	FMMY	TE	Cir Mhor
60031	FM	FMMY	TE	Ben Lui
60032	FC	FCHN	TO	William Booth
60033	FP	FPEK	CF	Anthony Ashley Cooper
60034	FM	FMEK	CF	Carnedd Llewelyn
60035	FM	FMEK	CF	Florence Nightingale
60036	FM	FMEK	CF	Sgurr Na Ciche
60037	FM	FMEK	CF	Helvellyn
60038	FM	FMMY	TE	Bidean Nam Bian
60039	FA	FASB	SL	Glastonbury Tor
60040	FA	FASB	SL	Brecon Beacons
60041	FA	FASB	SL	High Willhays
60042	FA	FASB	SL	Dunkery Beacon
60043	FA	FASB	SL	Yes Tor
60044	FM	FCHN	TO	Ailsa Craig
60045	FC	FCHN	TO	Josephine Butler
60046	FC	FCHN	TO	William Wilberforce
60047	FC	FCHN	TO	Robert Owen
60048	FA	FAXN	TO	Saddleback
60049	FM	FMMY	TE	Scafell
60050	FM	FPDI	IM	Roseberry Topping
60051	FP	FPDI	IM	Mary Somerville
60052	FM	FMMY	TE	Goat Fell
60053	FP	FPDI	IM	John Reith
60054	FP	FPDI	IM	Charles Babbage
60055	FC	FCHN	TO	Thomas Barnardo
60056	FC	FCHN	TO	William Beveridge
60057	FC	FCHN	TO	Adam Smith
60058	FC	FCHN	TO	John Howard
60059	FC	FCHN	TO	Samuel Plimsoll
60060	FC	FCHN	TO	James Watt
60061	FC	FCHN	TO	Alexander Graham Bell
60062	FP	FPEK	CF	Samuel Johnson
60063	FP	FPEK	CF	James Murray
60064	FP	FPDI	IM	Back Tor
60065	FP	FPEK	CF	Kinder Low
60066	FC	FCHN	TO	John Logie Baird
60067	FC	FCJN	TO	James Clerk-Maxwell
60068	FC	FCJN	TO	Charles Darwin
60069	FC	FCJN	TO	Humphry Davy
60070	FC	FCJN	TO	John Loudon McAdam
60071	FC	FCJN	TO	Dorothy Garrod
60072	FC	FCJN	TO	Cairn Toul
60073	FC	FCJN	TO	Cairn Gorm
60074	FC	FCJN	TO	Braeriach
60075	FC	FCJN	TO	Liathach
60076	FC	FCJN	TO	Suilven
60077	FC	FCJN	TO	Canisp
60078	FC	FCJN	TO	Stac Pollaidh
60079	FC	FCJN	TO	Foinaven
60080	FA	FABI	IM	Kinder Scout
60081	FA	FMEK	CF	Bleaklow Hill

60082	**FA**	FABI	IM	Mam Tor
60083	**FA**	FAXN	TO	Shining Tor
60084	**FA**	FABI	IM	Cross Fell
60085	**FA**	FABI	IM	Axe Edge
60086	**FC**	FCJN	TO	Schiehallion
60087	**FC**	FCJN	TO	Slioch
60088	**FC**	FCJN	TO	Buachaille Etive Mor
60089	**FC**	FCJN	TO	Arcuil
60090	**FC**	FPTY	TE	Quinag
60091	**FC**	FPDI	IM	An Teallach
60092	**FC**	FPEK	CF	Reginald Munns
60093	**FC**	FMEK	CF	Jack Stirk
60094	**FA**	FAXN	TO	Tryfan
60095	**FA**	FABI	IM	Crib Goch
60096	**FA**	FMEK	CF	Ben Macdui
60097	**FA**	FABI	IM	Pillar
60098	**FA**	FAXN	TO	Charles Francis Brush
60099	**FA**	FASB	SL	Ben More Assynt
60100	**FA**	FASB	SL	Boar of Badenoch

BR ELECTRIC LOCOMOTIVES

CLASS 73/0   ELECTRO – DIESEL   Bo – Bo

Built: 1962 by BR at Eastleigh Works.
Supply System: 660 – 850 V d.c. from third rail.
Engine: English Electric 4SRKT of 447 kW (600 hp) at 850 rpm.
Main Generator: English Electric 824/3D.
Traction Motors: English Electric 542A.
Max. Tractive Effort: Electric 187 kN (42000 lbf). Diesel 152 kN (34100 lbf).
Continuous Rating: Electric 1060 kW (1420 hp) giving a tractive effort of 43 kN (9600 lbf) at 55.5 mph.
Cont. Tractive Effort: Diesel 72 kN (16100 lbf) at 10 mph.
Maximum Rail Power: Electric 1830 kW (2450 hp) at 37 mph.

Brake Force: 31 t.	**Length over Buffers:** 16.36 m.
Design Speed: 80 mph.	**Weight:** 76.5 t.
Max. Speed: 60 mph.	**RA:** 6.
Wheel Diameter: 1016 mm.	**ETH Index (Elec. power):** 66

Train Brakes: Air, Vacuum and electro-pneumatic.
Multiple Working: Within sub-class, with Class 33/1 and various SR EMUs.
Communication Equipment: All equipped with driver – guard telephone.
Couplings: Drop-head buckeye.
Formerly numbered E 6001 – /3/5/6.

Non-standard Livery: 73005 is Network SouthEast blue.

73001		RCRB	BD
73002	**BR**	RCRB	BD
73003	**G**	NXXA	SL	Sir Herbert Walker
73005	**0**	RCRB	BD
73006	**BR**	RCRB	BD

CLASS 73/1 & 73/2   ELECTRO – DIESEL   Bo – Bo

Built: 1965 – 67 by English Electric Co. at Vulcan Foundry, Newton le Willows.
Engine: English Electric 4SRKT of 447 kW (600 hp) at 850 rpm.
Main Generator: English Electric 824/5D.
Traction Motors: English Electric 546/1B.
Supply System: 660 – 850 V d.c. from third rail.
Max. Tractive Effort: Electric 179 kN (40000 lbf). Diesel 160 kN (36000 lbf).
Continuous Rating: Electric 1060 kW (1420 hp) giving a tractive effort of 35 kN (7800 lbf) at 68 mph.
Cont. Tractive Effort: Diesel 60 kN (13600 lbf) at 11.5 mph.
Maximum Rail Power: Electric 2350 kW (3150 hp) at 42 mph.

Brake Force: 31 t.	**Length over Buffers:** 16.36 m.
Design Speed: 90 mph.	**Weight:** 77 t.
Max. Speed: 60 (90*) mph.	**RA:** 6.
Wheel Diameter: 1016 mm.	**ETH Index (Elec. power):** 66

Train Brakes: Air, Vacuum and electro-pneumatic.
Multiple Working: Within sub-class, with Class 33/1 and various SR EMUs.
Communication Equipment: All equipped with driver – guard telephone.

Couplings: Drop-head buckeye.
Non-standard Livery: 73101 is Pullman umber & Cream.

Class 73/2 are locos dedicated to InterCity and Network SouthEast services.
a Vacuum brake isolated.

Formerly numbered E 6001 – 20/22 – 26/28 – 49 (not in order).

73101		O	NKJL	SL	The Royal Alex'
73103		IO	NKJL	SL	
73104		IO	NKJL	SL	
73105		C	NKJL	SL	
73106		D	NKJL	SL	
73107		C	NKJL	SL	
73108		C	NKJL	SL	
73109	*	N	NWXB	BM	Battle of Britain 50th Anniversary
73110		C	NKJL	SL	
73112		N	IVGA	SL	University of Kent at Canterbury
73114		IO	NKJR	SL	
73117		IO	NKJL	SL	University of Surrey
73118		C	NKJL	SL	The Romney Hythe and Dymchurch Railway
73119		C	NKJL	SL	Kentish Mercury
73126		N	NKJR	SL	Kent & East Sussex Railway
73128		C	NKJR	SL	OVS BULLIED C.B.E. 1937 1949 C.M.E. SOUTHERN RAILWAY
73129		N	NKJL	SL	City of Winchester
73130		C	NKJL	SL	City of Portsmouth
73131		C	NKJL	SL	
73132		IO	NKJR	SL	
73133		N	NKJL	SL	The Bluebell Railway
73134		IO	NKJR	SL	Woking Homes 1885 – 1985
73136		N	NKJL	SL	Kent Youth Music
73138		C	NKJL	SL	
73139		IO	NKJR	SL	
73140		IO	NKJR	SL	
73141		IO	NKJR	SL	
73201 (73142)	a*	I	IVGA	SL	Broadlands
73202 (73137)	a*	I	IVGA	SL	Royal Observer Corps
73203 (73127)	a*	I	IVGA	SL	
73204 (73125)	a*	I	IVGA	SL	Stewarts Lane 1860 – 1985
73205 (73124)	*	M	IVGA	SL	London Chamber of Commerce
73206 (73123)	a*	I	IVGA	SL	Gatwick Express
73207 (73122)	a*	I	IVGA	SL	County of East Sussex
73208 (73121)	a*	I	IVGA	SL	Croydon 1883 – 1983
73209 (73120)	a*	I	IVGA	SL	
73210 (73116)	a*	I	IVGA	SL	Selhurst
73211 (73113)	a*	I	IVGA	SL	
73212 (73102)	a*	I	IVGA	SL	Airtour Suisse
73235 (73135)	a*	I	IVGA	SL	

NOTES FOR CLASSES 86 – 91.

The following common features apply to all locos of Classes 86 – 91.
Supply System: 25 kV a.c. from overhead equipment.
Communication Equipment: Driver – guard telephone and cab to shore radio-telephone.
Multiple Working: Time division multiplex system.

CLASS 86/1   BR DESIGN   Bo – Bo

Built: 1965 – 66 by English Electric Co. at Vulcan Foundry, Newton le Willows or BR at Doncaster Works. Rebuilt with Class 87 type bogies and motors. Tap changer control.
Traction Motors: GEC G412AZ frame mounted.
Max. Tractive Effort: 258 kN (58000 lbf).
Continuous Rating: 3730 kW (5000 hp) giving a tractive effort of 95 kN (21300 lbf) at 87 mph.
Maximum Rail Power: 5860 kW (7860 hp) at ?? mph.

Brake Force: 40 t.	**Length over Buffers:** 17.83 m.
Design Speed: 110 mph.	**Weight:** 87 t.
Max. Speed: 110 mph.	**RA:** 6.
ETH Index: 74	**Wheel Diameter:** 1150 mm.
Train Brakes: Air & Vacuum.	**Electric Brake:** Rheostatic.

Note: Class 86 were formerly numbered E 3101 – 3200 (not in order).

86101 (86201)	I	IWPA	WN	Sir William A Stanier FRS
86102 (86202)	I0	IWPA	WN	Robert A Riddles
86103 (86203)	I	IWPA	WN	André Chapelon

CLASS 86/2   BR DESIGN   Bo – Bo

Built: 1965 – 66 by English Electric Co. at Vulcan Foundry, Newton le Willows or BR at Doncaster Works. Later rebuilt with resilient wheels and flexicoil suspension. Tap changer control.
Traction Motors: AEI 282BZ.
Max. Tractive Effort: 207 kN (46500 lbf).
Continuous Rating: 3010 kW (4040 hp) giving a tractive effort of 85 kN (19200 lbf) at 77.5 mph.
Maximum Rail Power: 4550 kW (6100 hp) at 49.5 mph.

Brake Force: 40 t.	**Length over Buffers:** 17.83 m.
Design Speed: 125 mph.	**Weight:** 85 t – 86 t.
Max. Speed: 100 (110§) mph.	**RA:** 6.
ETH Index: 74	**Wheel Diameter:** 1156 mm.
Train Brakes: Air & Vacuum.	**Electric Brake:** Rheostatic.

86204	I	IWPA	WN	City of Carlisle
86205 (86503)	I	ICCA	LG	City of Lancaster
86206	M	ICCA	LG	City of Stoke on Trent
86207	I	IWPA	WN	City of Lichfield
86208	I	IWPA	WN	City of Chester

86209	§ M	IWPA	WN	City of Coventry
86210	I	IWPA	WN	City of Edinburgh
86212	M	ICCA	LG	Preston Guild 1328 – 1992
86213	I	IWPA	WN	Lancashire Witch
86214	I	ICCA	LG	Sans Pareil
86215	I	IANA	NC	Joseph Chamberlain
86216	I	ICCA	LG	Meteor
86217 (86504)	I	IANA	NC	Halley's Comet
86218	I	IANA	NC	Harold MacMillan
86219	I	IWPA	WN	Phoenix
86220	I	IANA	NC	The Round Tabler
86221	M	IANA	NC	B.B.C. Look East
86222 (86502)	I	ICCA	LG	LLOYD'S LIST
				250th ANNIVERSARY
86223	I	IANA	NC	Norwich Union
86224	§ I	IWPA	WN	Caledonian
86225	§ I	IWPA	WN	Hardwicke
86226	M	ICCA	LG	Royal Mail Midlands
86227	M	ICCA	LG	Sir Henry Johnson
86228	I	ICCA	LG	Vulcan Heritage
86229	I	ICCA	LG	Sir John Betjeman
86230	M	IANA	NC	The Duke of Wellington
86231	§ I	IWPA	WN	Starlight Express
86232	I	IANA	NC	Norwich Festival
86233 (86506)	I	ICCA	LG	Laurence Olivier
86234	I	ICCA	LG	J B Priestley OM
86235	I	IANA	NC	Crown Point
86236	I	IWPA	WN	Josiah Wedgwood
				MASTER POTTER 1736 – 1795
86237	I	IANA	NC	University of East Anglia
86238	I	IANA	NC	European Community
86239 (86507)	R	PXLE	CE	L S Lowry
86240	I	IWPA	WN	Bishop Eric Treacy
86241 (86508)	R	PXLE	CE	Glenfiddich
86242	M	IWPA	WN	James Kennedy GC
86243	RX	PXLE	CE	
86244	I	ICCA	LG	The Royal British Legion
86245	I	IWPA	WN	Dudley Castle
86246 (86505)	I	IANA	NC	Royal Anglian Regiment
86247	I	ICCA	LG	Abraham Darby
86248	I	IWPA	WN	Sir Clwyd – County of Clwyd
86249	M	IANA	NC	County of Merseyside
86250	I	IANA	NC	The Glasgow Herald
86251	I	IWPA	WN	The Birmingham Post
86252	I	ICCA	LG	The Liverpool Daily Post
86253 (86044)	I	IWPA	WN	The Manchester Guardian
86254 (86047)	RX	PXLE	CE	
86255 (86042)	I	ICCA	LG	Penrith Beacon
86256 (86040)	I	IWPA	WN	Pebble Mill
86257 (86043)	I	IWPA	WN	Snowdon
86258 (86501)	I	IWPA	WN	Talyllyn – The First Preserved
				Railway

86259	(86045)	**I** ICCA	LG	Peter Pan
86260	(86048)	**I** ICCA	LG	Driver Wallace Oakes G.C.
86261	(86041)	**RX** PXLE	CE	

CLASS 86/4 & 86/6   BR DESIGN   Bo–Bo

Built: 1965 – 66 by English Electric Co. at Vulcan Foundry, Newton le Willows or BR at Doncaster Works. Later rebuilt with resilient wheels and flexicoil suspension. Tap changer control.
Traction Motors: AEI 282AZ.
Max. Tractive Effort: 258 kN (58000 lbf).
Continuous Rating: 2680 kW (3600 hp) giving a tractive effort of 89 kN (20000 lbf) at 67 mph.
Maximum Rail Power: 4400 kW (5900 hp) at 38 mph.
Brake Force: 40 t.
Design Speed: 100 mph.
Max. Speed: 100 (75*) mph.
ETH Index: 74
Train Brakes: Air & Vacuum.
Length over Buffers: 17.83 m.
Weight: 83 t – 84 t.
RA: 6.
Wheel Diameter: 1156 mm.
Electric Brake: Rheostatic.

Class 86/6 have the ETH equipment isolated.

Note: 86405/11/4/5/28/31 have recently been renumbered from 86/6.

86401	(86001)	**RX** PXLE	CE	
86416	(86316)	**RX** PXLE	CE	
86417	(86317)	**RX** PXLE	CE	
86419	(86319)	**RX** PXLE	CE	Post Haste 150 YEARS OF TRAVELLING POST OFFICES
86424	(86324)	**R** PXLE	CE	
86425	(86325)	**R** PXLE	CE	
86426	(86326)	**RX** PXLE	CE	
86430	(86030)	**RX** PXLE	CE	
86602	(86402)	* **FD** MDNC	CE	
86603	(86403)	* **FD** MDNC	CE	
86604	(86404)	* **FD** MDNC	CE	
86605	(86405)	* **FD** MDNC	CE	Intercontainer
86606	(86406)	* **FD** MDNC	CE	
86607	(86407)	* **FD** MDNC	CE	The Institution of Electrical Engineers
86608	(86408)	* **FE** MDNC	CE	St. John Ambulance
86609	(86409)	* **FD** MDNC	CE	
86610	(86410)	* **FD** MDNC	CE	
86611	(86411)	* **FD** MDNC	CE	Airey Neave
86612	(86412)	* **FD** MDNC	CE	Elizabeth Garrett Anderson
86613	(86413)	* **FD** MDNC	CE	County of Lancashire
86614	(86414)	* **FD** MDNC	CE	Frank Hornby
86615	(86415)	* **FD** MDNC	CE	Rotary International
86618	(86418)	* **FD** MDNC	CE	
86620	(86420)	* **FD** MDNC	CE	
86621	(86421)	* **FD** MDNC	CE	London School of Economics
86622	(86422)	* **FD** MDNC	CE	
86623	(86423)	* **FD** MDNC	CE	

86627	(86427)	* FD	MDNC	CE	The Industrial Society
86628	(86428)	* FD	MDNC	CE	Aldaniti
86631	(86431)	* FD	MDNC	CE	
86632	(86432)	* FD	MDNC	CE	Brookside
86633	(86433)	* FD	MDNC	CE	Wulfruna
86634	(86434)	* FD	MDNC	CE	University of London
86635	(86435)	* FD	MDNC	CE	
86636	(86436)	* FD	MDNC	CE	
86637	(86437)	* FD	MDNC	CE	
86638	(86438)	* FD	MDNC	CE	
86639	(86439)	* FD	MDNC	CE	

CLASS 87　　　　　BR DESIGN　　　　Bo – Bo

Built: 1973 – 75 by BREL at Crewe Works. Class 87/1 has thyristor control instead of HT tap changing.
Traction Motors: GEC G412AZ frame mounted (87/0), G412BZ (87/1).
Max. Tractive Effort: 258 kN (58000 lbf).
Continuous Rating: 3730 kW (5000 hp) giving a tractive effort of 95 kN (21300 lbf) at 87 mph (Class 87/0), 3620 kW (4850 hp) giving a tractive effort of 96 kN (21600 lbf) at 84 mph (Class 87/1).
Maximum Rail Power: 5860 kW (7860 hp) at ?? mph.
Brake Force: 40 t.　　　　　　　**Length over Buffers:** 17.83 m.
Design Speed: 110 mph.　　　　**Weight:** 83.5 t.
Max. Speed: 110 (75*) mph.　　**RA:** 6.
ETH Index: 95　　　　　　　　　**Wheel Diameter:** 1150 mm.
Train Brakes: Air.　　　　　　　**Electric Brake:** Rheostatic.

Class 87/0. Standard Design. Tap Changer Control.

87001	I	IWCA	WN	Royal Scot
87002	I	IWCA	WN	Royal Sovereign
87003	I	IWCA	WN	Patriot
87004	I	IWCA	WN	Britannia
87005	I	IWCA	WN	City of London
87006	IO	IWCA	WN	City of Glasgow
87007	I	IWCA	WN	City of Manchester
87008	I	IWCA	WN	City of Liverpool
87009	I	IWCA	WN	City of Birmingham
87010	I	IWCA	WN	King Arthur
87011	I	IWCA	WN	The Black Prince
87012	M	IWCA	WN	The Royal Bank of Scotland
87013	I	IWCA	WN	John O' Gaunt
87014	I	IWCA	WN	Knight of the Thistle
87015	I	IWCA	WN	Howard of Effingham
87016	I	IWCA	WN	Willesden Intercity Depot
87017	IO	IWCA	WN	Iron Duke
87018	M	IWCA	WN	Lord Nelson
87019	IO	IWCA	WN	Sir Winston Churchill
87020	IO	IWCA	WN	North Briton
87021	I	IWCA	WN	Robert the Bruce
87022	M	IWCA	WN	Cock o' the North
87023	IO	IWCA	WN	Velocity

87024	I0	IWCA	WN	Lord of the Isles
87025	I0	IWCA	WN	County of Cheshire
87026	I	IWCA	WN	Sir Richard Arkwright
87027	I	IWCA	WN	Wolf of Badenoch
87028	I	IWCA	WN	Lord President
87029	I0	IWCA	WN	Earl Marischal
87030	I0	IWCA	WN	Black Douglas
87031	M	IWCA	WN	Hal o' the Wynd
87032	I0	IWCA	WN	Kenilworth
87033	M	IWCA	WN	Thane of Fife
87034	I0	IWCA	WN	William Shakespeare
87035	M	IWCA	WN	Robert Burns

Class 87/1. Thyristor Control.

| 87101 | * | FD | MDNC | CE | STEPHENSON |

CLASS 90   GEC DESIGN   Bo–Bo

Built: 1987 – 90 by BREL at Crewe Works. Thyristor control.
Traction Motors: GEC G412CY separately excited frame mounted.
Max. Tractive Effort: 192 kN (43150 lbf).
Continuous Rating: 3730 kW (5000 hp) giving a tractive effort of 95 kN (21300 lbf) at 87 mph.
Maximum Rail Power: 5860 kW (7860 hp) at ?? mph.
Brake Force: 40 t.   **Length over Buffers:** 18.80 m.
Design Speed: 110 mph.   **Weight:** 84.5 t.
Max. Speed: 110 (75*) mph.   **RA:** 7.
ETH Index: 95   **Wheel Diameter:** 1156 mm.
Train Brakes: Air.   **Electric Brake:** Rheostatic.
Couplings: Drop-head buckeye.

Non-standard Liveries:

90128 is in SNCB/NMBS (Belgian Railways) electric loco livery.
90129 is in DB (German Federal Railways) 'neurot' livery.
90130 is in SNCF (French Railways) 'Sybic' livery.
90136 is in livery "FD", but with yellow ends and roof.

Class 90/0. As built.

90001	I	IWCA	WN	BBC Midlands Today
90002	I	IWCA	WN	The Girls' Brigade
90003	I	IWCA	WN	
90004	I	IWCA	WN	The D' Oyly Carte Opera Company
90005	I	IWCA	WN	Financial Times
90006	I	IWCA	WN	High Sheriff
90007	I	IWCA	WN	Lord Stamp
90008	I	IWCA	WN	The Birmingham Royal Ballet
90009	I	IWCA	WN	Royal Show
90010	I	IWCA	WN	275 Railway Squadron (Volunteers)
90011	I	IWCA	WN	The Chartered Institute of Transport
90012	I	IWCA	WN	British Transport Police
90013	I	IWCA	WN	The Law Society
90014	I	IWCA	WN	'The Liverpool Phil'

90015	I	IWCA	WN	BBC North West
90016	RX	PXLA	CE	
90017	RX	PXLA	CE	
90018	RX	PXLA	CE	
90019	RX	PXLA	CE	Penny Black
90020	RX	PXLA	CE	Colonel Bill Cockburn CBE TD
90021	FD	PXLA	CE	
90022	FD	MDLC	CE	Freightconnection
90023	FD	MDLC	CE	
90024	FD	MDLC	CE	

Class 90/1. ETH equipment isolated. Renumbered from 90025 – 90050.

90125		FD	MDMC	CE	
90126	*	FD	MDMC	CE	Crewe Electric Depot Quality Approved
90127	*	FD	MDMC	CE	Allerton T&RS Depot Quality Approved
90128	*	O	MDMC	CE	Vrachtverbinding
90129	*	O	MDMC	CE	Frachtverbindungen
90130	*	O	MDMC	CE	Fretconnection
90131	*	M	MDMC	CE	
90132	*	FE	MDMC	CE	
90133	*	M	MDMC	CE	
90134	*	M	MDMC	CE	
90135	*	M	MDMC	CE	
90136	*	O	MDMC	CE	
90137	*	FD	MDMC	CE	
90138	*	FD	MDMC	CE	
90139	*	FD	MDMC	CE	
90140	*	FD	MDMC	CE	
90141	*	FD	MDMC	CE	
90142	*	FD	MDMC	CE	
90143	*	FD	MDMC	CE	
90144	*	FD	MDMC	CE	
90145	*	FD	MDMC	CE	
90146	*	FD	MDMC	CE	
90147	*	FD	MDMC	CE	
90148	*	FD	MDMC	CE	
90149	*	FD	MDMC	CE	
90150	*	FD	MDMC	CE	

CLASS 91   GEC DESIGN   Bo – Bo

Built: 1988 onwards by BREL at Crewe Works. Thyristor control.
Traction Motors: GEC G426AZ.
Continuous Rating: 4540 kW (6090 hp).
Maximum Rail Power: 4700 kW (6300 hp).
Brake Force: 45 t.   **Length over Buffers:** 19.40 m.
Design Speed: 140 mph.   **Weight:** 84 t.
Max. Speed: 140 mph.   **RA:** 7.
ETH Index: 95   **Wheel Diameter:** 1000 mm.
Train Brakes: Air.   **Electric Brake:** Rheostatic.
Couplings: Drop-head buckeye.

91001	I	IECA	BN	Swallow
91002	I	IECA	BN	Durham Cathedral
91003	I	IECA	BN	THE SCOTSMAN
91004	I	IECA	BN	The Red Arrows
91005	I	IECA	BN	Royal Air Force Regiment
91006	I	IECA	BN	
91007	I	IECA	BN	Ian Allan
91008	I	IECA	BN	Thomas Cook
91009	I	IECA	BN	Saint Nicholas
91010	I	IECA	BN	
91011	I	IECA	BN	Terence Cuneo
91012	I	IECA	BN	
91013	I	IECA	BN	Michael Faraday
91014	I	IECA	BN	Northern Electric
91015	I	IECA	BN	
91016	I	IECA	BN	
91017	I	IECA	BN	Commonwealth Institute
91018	I	IECA	BN	
91019	I	IECA	BN	Scottish Enterprise
91020	I	IECA	BN	
91021	I	IECA	BN	
91022	I	IECA	BN	Robert Adley
91023	I	IECA	BN	
91024	I	IECA	BN	
91025	I	IECA	BN	BBC Radio One FM
91026	I	IECA	BN	
91027	I	IECA	BN	
91028	I	IECA	BN	Guide Dog
91029	I	IECA	BN	Queen Elizabeth II
91030	I	IECA	BN	Palace of Holyroodhouse
91031	I	IECA	BN	Sir Henry Royce

CLASS 92 BRUSH DESIGN Co–Co

Built: 1993 onwards by Brush Traction. Thyristor control.
Supply System: 25 kV a.c. from overhead equipment and 750 V d.c. third rail.
Electrical equipment: ABB Transportation, Zürich, Switzerland.
Traction Motors: Brush.
Max. Tractive Effort: 400 kN (90 000 lbf).
Continuous Rating at Motor Shaft: 5040 kW (6760 hp).
Maximum Rail Power: 5000 kW (6700 hp).

Brake Force: t.	**Length over Buffers:** 21.34 m.
Design Speed: 140 mph.	**Weight:** 126 t.
Max. Speed: 140 km/h (87.5 mph).	**RA:** .
ETH Index:	**Wheel Diameter:** 1160 mm.
Train Brakes: Air.	**Electric Brake:** Rheostatic.

Couplings: Drop-head buckeye.
Multiple Working: Time division multiplex system.
Communication Equipment: Driver – guard telephone and cab to shore radio-telephone.
Cab Signalling: Fitted with TVM430 cab signalling for Channel Tunnel.

92001   FE
92002
92003
92004
92005
92006
92007
92008
92009
92010
92011
92012
92013
92014
92015
92016
92017
92018
92019
92020
92021
92022
92023
92024
92025
92026
92027
92028
92029
92030
92031
92032
92033
92034
92035
92036
92037
92038
92039
92040
92041
92042
92043
92044
92045
92046

SNCF CLASS 22200   B – B

Built: 1976 – 1986. by Alsthom/MTE. 22379/80 are converted from experimental locos 20011/2. These locos are included in this book becausse they will work to England under the Channel Tunnel.
Supply System: 25 kV a.c./1500 V d.c. from overhead equipment.
Traction Motors: Two Alsthom monomotors.
Max. Tractive Effort: 294 kN (66 150 lbf).
Continuous Rating at Motor Shaft: 4360 kW (5900 hp).
Maximum Rail Power: 5000 kW (6700 hp).

Brake Force: t.	**Length over Buffers:** 17.48 m.
Design Speed: 200 mph.	**Weight:** 89 t.
Max. Speed: 160 km/h (100 mph).	**RA:** .
ETH Index:	**Wheel Diameter:** 1250 mm.
Train Brakes: Air.	**Electric Brake:** Rheostatic.

Communication Equipment: Driver – guard telephone and cab to shore radio-telephone.
Cab Signalling: Fitted with TVM430 cab signalling for Channel Tunnel.

22379	(20011)	**0**	DP
22380	(20012)	**0**	DP
22399		**0**	DP	MORMANT
22400		**0**	DP	MONTIGNY-EN-OSTREVENT
22401		**0**	DP	MOULINS
22402		**0**	DP	SAINT-DIE-DES-VOSGES
22403		**0**	DP	NEUVES-MAISONS
22404		**0**	DP	PAVILLONS-SOUS-BOIS

Note: DP = Dijon Perrigny.

BR DEPARTMENTAL LOCOMOTIVES

CLASS 97/6   RUSTON SHUNTER   0-6-0

Built: 1959 by Ruston & Hornsby at Lincoln.
Engine: Ruston 6VPH of 123 kW (165 hp).
Main Generator: British Thomson Houston RTB6034.
Traction Motor: One British Thomson Houston RTA5041.
Max. Tractive Effort: 75 kN (17000 lbf).
Brake Force: 16 t.   **Length over Buffers:** 7.62 m.
Weight: 31 t.   **Wheel Diameter:** 978 mm.
Max. Speed: 20 mph.   **RA:** 1.
Train Brakes: Vacuum.

Non-Standard Livery: Departmental Yellow.

97651	(PWM 651)	v	**0**	REJK	CF
97654	(PWM 654)	v	**0**	IGJK	RG

CLASS 97/7   BATTERY LOCOS   Bo-Bo

Built: 1973 – 80 by BREL at Doncaster and Wolverton Works. Converted from Class 501 EMU cars.
Supply System: 750 V d.c. third rail or 320 V d.c. batteries.
Traction Motors: GEC WT344A.
Max. Tractive Effort: 73 kN (16400 lbf).
Brake Force: 45 t.   **Length over Buffers:** 18.44 m.
Weight: 59 t.   **Wheel Diameter:** 1071 mm.
Max. Speed: 25 mph.   **RA:** 4.
Multiple Working: Work in pairs.
Non-Standard Livery: Bright blue with yellow stripe.

97703	(61182)	a		NKFH	HE	97707	(61166)	a	**N**	NKFH	HE
97704	(61185)	a		NKFH	HE	97708	(61173)	a	**N**	NKFH	HE
97705	(61184)	a		NKFH	HE	97709	(61172)	a		NKFH	HE (S)
97706	(61189)	a		NKFH	HE	97710	(61175)	a		NKFH	HE (S)

CLASS 97/8   EE SHUNTER   0-6-0

For details see Class 09. Severn Tunnel emergency train locomotive.

Non-Standard Livery: BR blue with grey cab.

97806	(09017)	xo	**0**	IGJK	CF	Normally kept at Sudbrook

DB 968xxx SERIES

This number series was introduced in 1969 and is for former capital stock locomotives which no longer operate under their own power.

Non-standard Livery: 968021 is British Rail Research red/blue/white.

ADB 968021	(84009)	x	CE	Mobile load bank.
TDB 968030	(33018)	x		Moreton-in-Marsh training loco.

BR LOCOMOTIVES AWAITING DISPOSAL

03179N	Ryde T&RSMD	08677	Willesden TMD
08222	Bounds Green T&RSMD	08684	Bletchley TMD
08239	Neville Hill T&RSMD	08686	Allerton TMD
08254	Gateshead	08688	Allerton TMD
08305	Healey Mills	08699	Crewe Diesel TMD
08309F	Knottingley TMD	08707	Doncaster TMD
08390	Landore T&RSMD	08708	Colchester
08419	Kingmoor Yard	08719	Bletchley TMD
08427	March TMD	08733	Motherwell TMD
08434	Derby T&RSMD	08755	Millerhill
08439	Immingham TMD	08760	BRML Eastleigh
08468	Springs Branch	08771	Heaton T&RSMD
08473	Leicester	08777	Hull Botanic Gardens
08478	Immingham TMD	08778D	Cardiff Canton T&RSMD
08496	Cambridge T&RSMD	08787	ABB Crewe Works
08507	Reading T&RSMD	08789	Bletchley TMD
08508	Scunthorpe Yard	08793D	Aberdeen T&RSMD
08515	Gateshead	08794	Neville Hill T&RSMD
08518	March TMD	08797	Thornaby TMD
08521	Allerton TMD	08800I	Bristol Bath Road TMD
08532	Allerton TMD	08802	Heaton T&RSMD
08533	Colchester	08803D	Reading T&RSMD
08537FO	Bescot TMD	08804	Cardiff Canton T&RSMD
08544	Heaton T&RSMD	08814	Derby T&RSMD
08565	Motherwell TMD	08822	Cardiff Canton T&RSMD
08579	Hunslet Sidings	08829	Toton TMD
08583	Doncaster TMD	08831	Eastleigh T&RSMD
08589	Cardiff Canton T&RSMD	08836	Cardiff Canton T&RSMD
08590BS	Heaton T&RSMD	08838	Derby T&RSMD
08595	Doncaster TMD	08840	Allerton TMD
08604G	Derby T&RSMD	08841	ABB Crewe
08608	Gateshead	08848	Cardiff Canton T&RSMD
08609	Willesden TMD	08849	ABB Crewe
08614	Willesden TMD	08855	Aberdeen T&RSMD
08618	Gateshead	08858	Allerton TMD
08626	Allerton TMD	08859	March TMD
08631N	March TMD	08868	March TMD
08634	Stratford TMD	08870	Doncaster TMD
08638BS	Reading T&RSMD	08880	Tinsley TMD
08647G	BRML Doncaster	08885	Doncaster TMD
08656	Bletchley TMD	08889	March TMD
08657	York	08895	Landore T&RSMD
08658	Norwich T&RSMD	08898	Bescot TMD
08659	Healey Mills	08916	Allerton TMD
08660	Cardiff Canton T&RSMD	08917	Allerton TMD
08667	Neville Hill T&RSMD	08923F	Stratford TMD
08671	Gateshead	08929	Old Oak Common TMD
08672	Bescot TMD	08935	Bristol Bath Road TMD

08936	March TMD
08949	Bristol Bath Road TMD
20009	Thornaby TMD
20010 FR	Falkland Junction
20011	Derby T&RSMD
20019	Falkland Junction
20025	Scunthorpe Yard
20028 BS	Falkland Junction
20042	Scunthorpe Yard
20043	Scunthorpe Yard
20055	Falkland Junction
20058	Falkland Junction
20061	Scunthorpe Yard
20068	Immingham TMD
20071	Falkland Junction
20082	Falkland Junction
20089	Immingham TMD
20090 FR	Falkland Junction
20094	Thornaby TMD
20096	Thornaby TMD
20106	BRML Doncaster
20119	Toton
20135	Toton
20142	Toton
20151	Falkland Junction
20160 0	Bescot Yard
20172	Falkland Junction
20176	Scunthorpe Yard
20177	Toton TMD
20181	Bescot Yard
20185	Falkland Junction
20186	Falkland Junction
20195	Falkland Junction
20197	Falkland Junction
20220	Kingmoor Yard
20221	Kingmoor Yard
20223	Kingmoor Yard
25194	Bescot Yard
25205	Bescot Yard
25206	BRML Doncaster
25211	Bescot Yard
25259	Bescot Yard
26001 G	Inverness T&RSMD
26002 FC	Inverness T&RSMD
26003 C	Inverness T&RSMD
26004 C	Inverness T&RSMD
26005 C	Inverness T&RSMD
26006 FC	Inverness T&RSMD
26007 G	Inverness T&RSMD
26008 C	Inverness T&RSMD
26010 FR	Inverness T&RSMD
26011 C	Motherwell TMD
26014	Perth
26015	Inverness T&RSMD
26021	Inverness T&RSMD
26024	Motherwell TMD
26025 C	Inverness T&RSMD
26026 C	Perth
26027	Perth
26032 FR	Inverness T&RSMD
26035 C	Inverness T&RSMD
26036 C	Inverness T&RSMD
26037 FR	Inverness T&RSMD
26038 FR	Inverness T&RSMD
26040 C	Perth
26041 FR	Inverness T&RSMD
26042	Inverness T&RSMD
26043 C	Perth
31101 0	Bescot Yard
31108 FO	Scunthorpe Yard
31123	Bescot Yard
31156	Scunthorpe Yard
31162	Immingham TMD
31168	Bescot Yard
31196 C	Stratford TMD
31210 FO	Scunthorpe Yard
31212	Scunthorpe Yard
31215 FO	Immingham TMD
31217 FC	Crewe Diesel TMD
31221	Scunthorpe Yard
31223	Ripple Lane
31234 FO	Bescot TMD
31240 FO	Stratford TMD
31243 FO	Stratford TMD
31249	Scunthorpe Yard
31264	Thornaby TMD
31283 0	Stratford TMD
31286	Bescot Yard
31289	Bescot Yard
31293	Stratford TMD
31296 FA	Crewe Diesel TMD
31299 FO	Stratford TMD
31305	Bescot Yard
31320	Stratford TMD
31402	Bescot Yard
31404	BRML Doncaster
31428	Basford Hall Yard
31442	Crewe Diesel TMD
31460	Bescot TMD
31970 0	ABB Crewe
33006	BRML Eastleigh
33009 C	BRML Eastleigh
33020	Stewarts Lane T&RSMD
33029	Stewarts Lane T&RSMD

33033 FA	Stewarts Lane T&RSMD	47118 BR	Healey Mills
33038	Stratford TMD	47119 FP	Immingham TMD
33040	Stewarts Lane T&RSMD	47120 BR	Doncaster Belmont Yard
33047 C	Eastleigh T&RSMD	47123	Doncaster TMD
33050 FA	Stewarts Lane T&RSMD	47143	Doncaster TMD
33058	BRML Eastleigh	47195	Tinsley TMD
33101 D	Eastleigh T&RSMD	47198	Cardiff Canton T&RSMD
33103 C	Eastleigh T&RSMD	47199	Kingmoor Yard
33108 C	Eastleigh T&RSMD	47220 FO	Tinsley TMD
33110	Eastleigh T&RSMD	47227 FR	Tinsley TMD
33113	Stewarts Lane T&RSMD	47233 FP	Scunthorpe Yard
33114 N	Eastleigh T&RSMD	47318 FO	Bescot TMD
33117	Stewarts Lane T&RSMD	47320 FO	ABB Crewe
33118 C	Eastleigh T&RSMD	47324 FP	Immingham TMD
33201 C	Stewarts Lane T&RSMD	47373 FP	Scunthorpe Yard
33205 FD	Stewarts Lane TMD	47380 FP	Scunthorpe Yard
33211 FD	Stewarts Lane T&RSMD	47381 FP	Scunthorpe Yard
37008 FR	Tinsley TMD	47406 IO	Scunthorpe Yard
37190 FM	Gateshead	47407 BR	Scunthorpe Yard
37215 FP	Inverness T&RSMD	47411 BR	Scunthorpe Yard
37681 FA	ABB Crewe	47413 BR	Scunthorpe Yard
45013	March Whitemoor Yard	47417	Scunthorpe Yard
45015	Toton TMD	47418	Scunthorpe Yard
45041	Thornaby TMD	47421	Crewe Diesel TMD
45058	March Whitemoor Yard	47423	Old Oak Common TMD
45062	March Whitemoor Yard	47424 BR	ABB Crewe
45076	March Whitemoor Yard	47425	Old Oak Common TMD
45114	March Whitemoor Yard	47426 BR	Old Oak Common TMD
45119	March Whitemoor Yard	47430 FA	Old Oak Common TMD
45122	March Whitemoor Yard	47431 BR	Old Oak Common TMD
45127	March Whitemoor Yard	47432 BR	Gresty Lane
45137	March Whitemoor Yard	47433 BR	Crewe Diesel TMD
45139	March Whitemoor Yard	47438 BR	Old Oak Common TMD
45142	March Whitemoor Yard	47439 BR	Crewe Diesel TMD
45143	March Whitemoor Yard	47440 BR	Old Oak Common TMD
46010	Doncaster TMD	47441 BR	Old Oak Common TMD
46023	Basford Hall Yard	47442 BR	Crewe Diesel TMD
47002	Doncaster Belmont Yard X	47443 BR	Crewe Diesel TMD
47007 FA	Doncaster Belmont Yard X	47444 BR	Basford Hall Yard
47008	Stratford TMD	47445 FD	Doncaster Belmont Yard X
47011	ABB Crewe	47446 BR	Old Oak Common TMD
47018 FO	Doncaster TMD	47447 BR	Doncaster Belmont Yard
47094 FP	Scunthorpe Yard	47448 BR	Holbeck
47096	Tinsley TMD	47451 BR	Doncaster Belmont Yard X
47099 FO	Doncaster Belmont Yard X	47452 BR	Old Oak Common TMD
47100	Doncaster Belmont Yard X	47453 BR	Old Oak Common TMD
47102	Tinsley TMD	47454 BR	Doncaster Belmont Yard
47107 FO	Doncaster Belmont Yard	47455 BR	ABB Crewe
47112 FO	Old Oak Common TMD	47457 BR	Old Oak Common TMD
47115	Scunthorpe Yard	47458 R	Crewe Diesel TMD
47116	ABB Crewe	47465 BR	Old Oak Common TMD
47117	Doncaster Belmont Yard	47466 BR	Holbeck

47470 **M**	ABB Crewe	47643 **IO**	Inverness T&RSMD
47472	Old Oak Common TMD	50029 **N**	Laira T&RSMD
47482 **BR**	Crewe Diesel TMD	50030 **N**	Laira T&RSMD
47483 **M**	Crewe Diesel TMD	56002 **FC**	Doncaster TMD
47485 **BR**	Crewe Diesel TMD	56015 **FC**	Toton TMD
47488 **BR**	Crewe Diesel TMD	56017 **FC**	Toton TMD
47509 **I**	Bristol Bath Road TMD	56028 **FC**	Toton TMD
47515 **M**	Holbeck	56030 **FC**	Toton TMD
47518 **BR**	Immingham TMD	56042 **F**	Toton TMD
47527 **M**	Bristol Bath Road TMD	56122 **FC**	Toton TMD
47533 **R**	Old Oak Common TMD	73004 **0**	Stewarts Lane T&RSMD
47534 **BR**	ABB Crewe	73111 **IO**	Stewarts Lane T&RSMD
47538 **BR**	ABB Crewe	97250 **0**	Inverness T&RSMD
47539 **RX**	Crewe Diesel TMD	97251 **M**	Inverness T&RSMD
47542	Stratford TMD	97252 **M**	Inverness T&RSMD
47549 **IO**	Crewe Diesel TMD	97653 **0**	Reading T&RSMD
47585 **BR**	Holbeck	97701 **0**	Birkenhead N. T&RSMD
47633 **BR**	BRML Glasgow	97702 **0**	Birkenhead N. T&RSMD

Non-Standard Liveries:

08761 Provincial grey light blue, white and dark blue
20160 BR blue but with red cab & silver roof
31101 BR blue but with large logo and red lower bodyside stripe.
31283 BR blue with large numbers
31970 Research light grey, dark grey, white and red
73004 NSE blue
97250 BR carriage blue & grey
97653 Departmental yellow
97701 Bright blue with yellow stripe
97702 Bright blue with yellow stripe

Note: 97701/2 carry DB 977363/2 in error.

Class 47 No. 47307 in the new Railfreight Distribution 'European' livery
pictured about to join the North London line at Stratford on 19th October
93. *Norman Barrington*

Class 50s Nos. 50033 'Glorious', 50050 & 50007 'SIR EDWARD ELGAR'
ole-head the 'Bishops Triple' at Whittington near Chesterfield on 5th June
93. *Peter Fox*

Revised blue liveried Class 47 No. 47971 'Robin Hood' hauls the diverted 07.13 Carlisle – London Euston away

Class 56 No. 56032, in Trainload Metals livery, provides the power for a ...rt steel train passing through Sonning cutting on 4th April 1992.

Hugh Ballantyne

Foster-Yeoman owned Class 59 No. 59004 'YEOMAN CHALLENGER' ...nts hoppers in Acton yard on 26th March 1991. *Norman Barrington*

Trainload Coal liveried Class 58 No. 58042 'Ironbridge Power Station' with a rake of empty HAAs at Burton

Class 60 No. 60080 'Kinder Scout' in Trainload construction livery at Melton with the 17.35 Hull – Rylstone on 17th August 1993.

Ian A. Lyall

Class 73 No. 73134 'W.Walkes' passes through the Frome valley, near 1995.

Parcels red liveried Class 86 No. 86425 is pictured passing Low Gill with 07 Liverpool – Edinburgh on 13th March 1993.   *Hugh Ballantyne*

Unique Class 87/1 No. 87101 'STEPHENSON' in Railfreight Distribution livery ads south with an Aberdeen – Willesden Brent sidings tank train on 12th ne 1992.   *Paul Senior*

▲ Ex-works Rail Express Systems liveried Class 90 No. 90019 'Penny Bla
passes Norton Bridge on 9th April 1992 with the 06.30 Glasgow – Eust
Downside carriage empty NPCCS. *Hugh Ballanty*

▼ Class 91 No. 91028 'Guide Dog' is pictured at Newcastle on 28th Aug
1993. *A.O. Wy*

PLATFORM 5
EUROPEAN RAILWAY HANDBOOKS

The Platform 5 European Railway Handbooks are the most comprehensive guides to the rolling stock of selected European railway administrations available. Each book lists all locomotives and multiple units of the country concerned, giving details of number carried, livery and depot allocation, together with a wealth of technical data for each class of vehicle. Lists of preserved locos and MUs are also included, plus a guide to preservation centres. Each book is A5 size, thread sewn and includes at least 16 pages of colour photographs.

The new third edition of German Railways Locomotives & Multiple Units and both French Railways/Chemins de Fer Francais and Swiss Railways/Chemins de fer Suisses, contain 32 pages of colour illustrations. French Railways and Swiss Railways are produced in English and French. The full range of overseas titles available is as follows:

German Railways 3rd edition. £12.50
French Railways/Chemins de Fer Francais 2nd ed. . .£9.95
Swiss Railways/Chemins de fer Suisses. £9.95
ÖBB/Austrian Federal Railways 2nd edition. £6.95
Benelux Locomotives & Coaching Stock 2nd ed. . . .£6.95
A Guide to Portuguese Railways (Fearless). £4.95

Other Publishers' Overseas Titles:

Canadian Trackside Guide 1990 (Bytown). £9.95
TGV Handbook (Capital). £7.95
Paris Metro Handbook (Capital). £7.95
World Metro Systems (Capital). £7.95

All these publications are available from shops, bookstalls or direct from: Mail Order Department, Platform 5 Publishing Ltd., Wyvern House, Sark Road, SHEFFIELD, S2 4HG, ENGLAND. For a full list of titles available by mail order, please send SAE to the above address.

POOL CODES & ALLOCATIONS

CENTRAL SERVICES

CDJB   Research. BS Class 20.

20087   20092 **CS** 20118 **FR** 20132 **FR** 20165 **FR** 20169 **CS**

CDJC   Research. BS Class 47.

47971 **BR** 47972 **CS** 47973 **M** 47975 **C** 47976 **C** 47981 **C**

CEJB   Civil Link. BS Class 31.

31102 **C** 31105 **C** 31106 **C** 31107 **C** 31110 **C** 31112 **C** 31113 **C**
31405 **M** 31415   31462 **D** 31467

CEJC   Civil Link. BS Class 47.

47300 **C** 47332 **C** 47333 **C** 47341 **C** 47353 **C** 47356 **FO** 47357 **C**
47372 **C**

CEJX   Civil Link. Stored.

20066   20138 **FR** 31125 **C**

TRAINLOAD FREIGHT SECTOR

FABI   Construction. IM (based at Buxton).

37405 **M** 37417 **M** 37420 **M** 37520 **FM** 37677 **FA** 37686 **FA** 60005 **FA**
60015 **FA** 60016 **FA** 60080 **FA** 60082 **FA** 60084 **FA** 60085 **FA** 60095 **FA**
60097 **FA**

FASB   Construction. SL.

56033 **FA** 56035 **FA** 56037 **FA** 56041 **FA** 56051 **FA** 56055 **FA** 56056 **FA**
56057 **FA** 56058 **FA** 56059 **FA** 56065 **FA** 56070 **FA** 56103 **FA** 56105 **FA**
60001 **FA** 60017 **FA** 60018 **FA** 60019 **FA** 60039 **FA** 60040 **FA** 60041 **FA**
60042 **FA** 60043 **FA** 60099 **FA** 60100 **FA**

FAXN   Construction. TO (based at Leicester).

60006 **FA** 60009 **FA** 60010 **FA** 60011 **FA** 60012 **FA** 60048 **FA** 60083 **FA**
60094 **FA** 60098 **FA**

FCBN   Power Station Coal. TO Classes 56 & 58 (East Midlands).

56004   56007 **FC** 56009 **FC** 56010   58001 **FC** 58002 **FC** 58003 **FC**
58004 **FC** 58005 **FC** 58006 **FC** 58007 **FC** 58008 **FC** 58009 **FC** 58010 **FC**
58011 **FC** 58012 **FC** 58013 **FC** 58014 **FC** 58015 **FC** 58016 **FC** 58017 **FC**
58018 **FC** 58019 **FC** 58020 **FC** 58021 **FC** 58022 **FC** 58023 **FC** 58024 **FC**
58025 **FC** 58026 **FC** 58027 **FC** 58028 **FC** 58029 **FC** 58030 **FC** 58031 **FC**
58032 **FC** 58033 **FC** 58034 **FC** 58035 **FC** 58036 **FC** 58037 **FC** 58038 **FC**
58041 **FC** 58050 **FC**

FCCI   Coal. IM.

56003 **F** 56085 **FC** 56088 **FC** 56089 **FC** 56090 **FC**

CDN   Power Station Coal. TO (Yorkshire).

```
6005 FC 56006 FC 56011 FR 56021 FC 56067 FC 56068 FC 56071 FC
6072 F  56074 FC 56075 FC 56077 FC 56078 FC 56079 FC 56080 FC
6082 FC 56083 FC 56086 FC 56091 FC 56093 FC 56095 FC 56096 FC
6098 FC 56100 FC 56101 FC 56102 FC 58039 FC 58040 FC 58042 FC
8043 FC 58044 FC 58045 FC 58046 FC 58047 FC 58048 FC 58049 FC
```

CEN   Power Station Coal. TO (North East).

```
6062 FA 56081 FC 56107 FC 56108 FR 56109 FC 56110 FA 56111 FC
6112 FC 56117 FC 56118 FC 56120 FC 56130 FC 56131 FC 56133 FC
6134 FC 56135 F
```

CFN   Nuclear Flask Traffic. TO.

```
1130 FC 31199 FC 31200 FC 31201 FC 31224 C  31275 FC 31302 FP
1304 FC 31312 FC 31319 FC
```

CHN   Power Station Coal. TO Class 60 (North West).

```
0032 FC 60044 FM 60045 FC 60046 FC 60047 FC 60055 FC 60056 FC
0057 FC 60058 FM 60060 FC 60061 FC 60066 FC
```

CJN   Power Station Coal. TO Class 60 (East Midlands).

```
0004 FC 60059 FC 60067 FC 60068 FC 60069 FC 60070 FC 60071 FC
0072 FC 60073 FC 60074 FC 60075 FC 60076 FC 60077 FC 60078 FC
0079 FC 60086 FC 60087 FC 60088 FC 60089 FC
```

CKK   Power Station Coal. CF.

```
7701 FC 37702 FC 37703 FC 37704 FC 37796 FC 37797 FC 37799 FC
7802 FC 37887 FC 37889 FC 37894 FC 37895 FC 37896 FC 37897 FC
7898 FC 37899 FC 56113 FC 56114 FC 56115 FC 56119 FC
```

CNN   Power Station Coal. TO Class 56 (North West).

```
6018 FC 56019 FR 56029 FC 56092 F  56099 FC 56132 FC
```

CPA   Coal & Petroleum. TO (Scotland).

```
6025 FC 56104 FC 56121 FC 56123 FC 56124 FC 56125 FC 56127 FC
6128 FC 56129 FC
```

CPM   Coal & Petroleum. ML (Scotland).

```
7051 FM 37066 C  37071 C  37100 FM 37111 FM 37116 BR 37184 C
7188 C  37212 FC 37403 FD 37404 M  37690 FO 37692 FC 37693 FC
7695 FC 37696 FC 37714 FM
```

CXX   Power Station Coal. Stored.

```
0016    20057    20059 FR 20073    20081    20154    20168
6008    56012 FC 56013 FC 56014 FC 56016 FC 56020    56022
6023 FC 56024 FO 56026    56027 FC 56066 FC
```

MCK   Metals. CF Class 56.

```
6032 FM 56038 FM 56040 FM 56044 FM 56053 FA 56054 FM 56060 FM
6064 FM 56073 FM 56076 FM
```

FMEK Metals. CF Class 60.

60029 FM 60034 FM 60035 FM 60036 FM 60037 FM 60081 FA 60093 P 60096 FA

FMHK Metals. CF Class 37.

37901 FM 37902 FM 37903 FM 37904 FM 37905 FM 37906 FM

FMMY Metals. TE Class 60.

60020 FM 60022 FM 60023 FM 60030 FM 60031 FM 60038 FM 60049 P 60052 FM

FMPY Metals. TE Class 37.

37415 M 37419 M 37426 M 37514 FM 37516 FM 37682 FA

FMTY Metals. TE Class 56.

56034 FA 56039 FA 56043 FA 56045 FA 56050 FA 56061 FM 56063 P 56069 FM 56087 FM 56097 FM 56116 FC

FMXX Metals. Stored.

20137 FR 56052 FM

FPCI Petroleum. IM Class 37.

37688 FA 37698 FC 37699 FC 37706 FP 37707 FP 37708 FP 37711 P 37713 FM 37715 FP 37717 FP 37719 FP 37798 FC 37800 FC 37803 P 37883 FM 37884 F 37885 FM 37886 FM 37891 FP

FPDI Petroleum. IM Class 60.

60002 FP 60003 FP 60008 FM 60013 FP 60014 FP 60021 FM 60024 P 60025 FP 60026 FP 60027 FP 60028 FP 60050 FM 60051 FP 60053 P 60054 FP 60064 FP 60091 FC

FPEK Petroleum. CF.

37521 FP 37668 FP 60033 FP 60062 FP 60063 FP 60065 FP 60092 P

FPFR Petroleum. IM (based at Ripple Lane).

37667 FP 37676 FA 37678 FA 37679 FA 37705 FP 37709 FP 37710 P 37890 FP 37892 FP

FPGI Petroleum. IM Class 56.

56084 FC 56094 FC 56106 FC 56126 FC

FPGM Petroleum. ML.

37712 FP 37801 FC 37893 FP

FPJI Petroleum. IM Infrastructure Class 37.

37501 FM 37513 FM 37515 FM 37517 FM 37519 FM 37694 FC

FPJW Petroleum. IM Infrastructure Class 37 (weekend work).

37009 FD 37502 FM 37512 FM 37687 FA

PRI Petroleum. IM (for contract services).

37350 FP 37358 F 37359 FP 37378 FD 37508 FM 37680 FA 37684 FA
37689 FC 37691 FO

PTY Petroleum. TE.

37506 FM 37697 FC 37716 FM 37718 FM 60007 FP 60090 FC

PYI Petroleum. IM (restricted use).

37382 FP 37511 FM

PYX Petroleum. Stored.

37381 FD 37507 FM 37888 FP

QXA Railfreight Headquarters. Stored Class 37.

37004 FM 37031 FD 37037 FM 37046 C 37092 C 37101 FD 37223 FC
37229 FC 37235 F 37518 FM

QXB Railfreight Headquarters. Stored Class 56.

56001 FA

SCD Doncaster Shunters.

08418 F 08442 F 08500 O 08512 F 08514   08682   08813 D
08824 F 08877 D 08903

SCK Knottingley Shunters.

08499 F 08516 D 08525 F 08605   08662   08706   08776 D
08782   08783   08806 F 09005 D 09014 D

SCN Toton Shunters.

08441   08449   08492   08511   08597   08607   08623
08723   08773   09104 D 09201 D

SNH Heaton Shunters.

08578 R 08587   08701 R 08886   08888 R 08931

SNI Immingham Shunters.

08388 F 08401 D 08405 D 08466 FO 08632   08665

SNL Neville Hill Shunters.

08575 BS 08588 BS 08745 BS 08908   08950 I

SNT Tinsley Shunters.

08389   08509 F 08510   08581   08691 G 08879

SNX North East Shunters. Spares & Stored.

08445   08577   08661   08919

SNY Thornaby Shunters.

08582 D 09106 D 09204 D

SSA Ayr Shunters.

08561   08586 F 08675 F

FSSB   Motherwell Shunters (based at Aberdeen).

08882   09205 **D**

FSSI   Inverness Shunters.

08754   08762

FSSM   Motherwell Shunters.

08411   08506   08568   08571   08622   08630   08693
08718   08720 **D** 08730 **O** 08731   08735   08738 **D** 08853
08881 **D** 08883 **O** 08906   08922 **D** 08952   09103 **D** 09202 **D**

FSSX   Scottish Shunters. Spares & Stored.

08938 **O**

FSWK   Cardiff Canton Shunters.

08493   08646 **F** 08756 **D** 08770 **D** 08786 **D** 08798   08830
08896   08932   08942   09001   09008 **D** 09013 **D** 09015 **D**
09105 **D** 09107 **D** 09203 **D**

FSWL   Landore Shunters.

08993   08994 **FR** 08995 **FC**

FSWX   Western Shunters. Spares & Stored.

08481   08664   08780   08795 **D** 08867 **O**

INTERCITY SECTOR

IANA   Anglia Services Locos.

86215 **I** 86217 **I** 86218 **I** 86220 **I** 86221 **M** 86223 **I** 86230 **M**
86232 **I** 86235 **I** 86237 **I** 86238 **I** 86246 **I** 86249 **M** 86250 **I**

IBRA   Bristol Class 47/4. Special Use.

47809 **I** 47819 **I** 47820 **I** 47821 **I** 47823 **I** 47833 **G** 47834 **I**
47835 **I**

IBRB   Bristol Class 47/4. General Hire.

47840 **I** 47843 **I** 47851 **I**

ICCA   Cross Country Services Class 86.

86205 **I** 86206 **M** 86212 **M** 86214 **I** 86216 **I** 86222 **I** 86226 **M**
86227 **M** 86228 **I** 86229 **I** 86233 **I** 86234 **I** 86244 **I** 86247 **I**
86252 **I** 86255 **I** 86259 **I** 86260 **I**

ICCP   Cross Country Services Class 43. LA & PM.

43086 **I** 43087 **I** 43088 **I** 43089 **I** 43101 **I** 43102 **I** 43103 **I**
43121 **I** 43122 **I** 43153 **I** 43154 **I** 43155 **I** 43156 **I** 43157 **I**
43158 **I** 43159 **I** 43160 **I** 43161 **I** 43162 **I** 43170 **I** 43180 **I**
43193 **I** 43194 **I** 43195 **I** 43196 **I** 43197 **I** 43198 **I**

CCS   Cross Country Services Class 43. EC.

43013 I	43014 I	43062 I	43063 I	43065 I	43067 I	43068 I
43069 I	43070 I	43071 I	43078 I	43079 I	43080 I	43084 I
43090 I	43091 I	43092 I	43093 I	43094 I	43097 I	43098 I
43099 I	43100 I	43123 I				

ECA   ECML Services Class 91.

91001 I	91002 I	91003 I	91004 I	91005 I	91006 I	91007 I
91008 I	91009 I	91010 I	91011 I	91012 I	91013 I	91014 I
91015 I	91016 I	91017 I	91018 I	91019 I	91020 I	91021 I
91022 I	91023 I	91024 I	91025 I	91026 I	91027 I	91028 I
91029 I	91030 I	91031 I				

ECP   ECML Services Class 43.

43038 I	43039 I	43095 I	43096 I	43104 I	43105 I	43106 I
43107 I	43108 I	43109 I	43110 I	43111 I	43112 I	43113 I
43114 I	43115 I	43116 I	43117 I	43118 I	43119 I	43120 I

EJI   Infrastructure. ECML. IM.

31531 C	31541 C	31544 C	31549 C	31552 C	31556 C	31558 C
37003 C	37058 C	37095 C	37104 C	47331 C	47676 I	

EJT   Infrastructure. ECML. IM (ECML standby locomotives).

47520 **M** 47671 **BR** 47673 **IO** 47675 **M**

EJW   Infrastructure. ECML. IM (weekend work).

31149 **FR** 31184 **FO** 31205 **FR** 31276 **FC** 31294 **FA** 31547 **C** 31553 **C**
37271 **FD** 47319 **FP** 47550 **M**

GJA   Infrastructure. GWML BR (weekend work).

37040 **FM** 37042 **FM** 37048 **FM** 37065 **FD** 37072 **D** 37074 **FD** 37077 **FM**
37109 **FM** 37137 **FM** 37138 **FM** 37203 **FM** 37213 **FC** 37219   37222 **FC**
37227 **FM**

GJK   Infrastructure. GWML. BR.

37010 **C** 37012 **C** 37035 **C** 37038 **C** 37054 **C** 37097 **C** 37098 **C**
37174 **C** 37264 **C** 37372 **C**

GJO   Infrastructure. GWML. OC.

47004 **G** 47016 **FO** 47108   47121   47315 **C** 47366 **FO** 47484 **G**

HFB   M&EE. Overhead Line Maintenance. BS.

31271 **FA** 31403   31407 **M** 31435 **C** 31459   31461 **D** 31466 **C**

ISA   Inverness Class 37.

37078 **FM** 37080 **FP** 37113 **FD** 37133 **C** 37152 **I** 37170 **C** 37175 **C**
37214 **FA** 37221 **I** 37250 **FM** 37251 **I** 37262 **D** 37505 **I** 37510 **I**
37683 **I** 37685 **FR**

ILRA   Bristol Class 47/4. Extended Range Locos.

47805 I	47806 I	47807 I	47810 I	47811 I	47812 I	47813 I
47814 I	47815 I	47816 I	47817 I	47818 I	47822 I	47825 M
47826 I	47827 I	47828 I	47829 I	47830 I	47831 M	47832 M
47839 I	47841 I	47844 I	47845 I	47846 I	47847 I	47848 I
47849 M	47850 I	47853 M				

IMJB   Infrastructure. Midland/Cross Country. BS.

31116 0

IMJC   Infrastructure. Midland/Cross Country. BR.

47348 FO 47368 FP 47462 R 47555 IO 47674 BR 47677 I 47802 I
47803 O 47804 I

IMJW   Infrastructure. Midland/Cross Country. BS (weekend work).

31119 C 31126 C 31134 C 31145 C 31209 FA 31252 FO 31411 D
31418   31427

IMLP   MML Services Class 43.

43043 I	43044 I	43045 I	43046 I	43047 I	43048 I	43049 I
43050 I	43051 I	43052 I	43053 I	43054 I	43055 I	43056 I
43057 I	43058 I	43059 I	43060 I	43061 I	43064 I	43066 I
43072 I	43073 I	43074 I	43075 I	43076 I	43077 I	43081 I
43082 I	43083 I	43085 I				

IVGA   Gatwick Express Services Class 73.

| 73112 N | 73201 I | 73202 I | 73203 I | 73204 I | 73205 M | 73206 I |
| 73207 I | 73208 I | 73209 I | 73210 I | 73211 I | 73212 I | 73235 I |

IWCA   WCML Services Classes 87 & 90.

87001 I	87002 I	87003 I	87004 I	87005 I	87006 IO	87007 I
87008 I	87009 I	87010 I	87011 I	87012 M	87013 I	87014 I
87015 I	87016 I	87017 IO	87018 M	87019 IO	87020 IO	87021 I
87022 M	87023 IO	87024 IO	87025 IO	87026 I	87027 I	87028 I
87029 IO	87030 IO	87031 M	87032 IO	87033 M	87034 IO	87035 I
90001 I	90002 I	90003 I	90004 I	90005 I	90006 I	90007 I
90008 I	90009 I	90010 I	90011 I	90012 I	90013 I	90014 I
90015 I						

IWCP   WCML Services Class 43.

| 43042 I | 43141 I | 43142 I | 43144 I | 43145 I | 43146 I | 43147 I |

IWJB   Infrastructure. WCML. BS.

31417 D	31420 M	31422 M	31423 M	31434	31450	31457
31516 C	31524 C	31530 C	31537 C	31545	31546 C	31551 I
31554 C						

IWJC   Infrastructure. WCML. CD Class 31.

| 31142 C | 31144 C | 31154 C | 31159 C | 31163 C | 31203 C | 31235 I |
| 31512 C | 31519 C | | | | | |

VJD   Infrastructure. WCML. CD Class 47.

7329 **C** 47334 **C** 47340 **C** 47473 **BR** 47478   47525 **IO**

VJI   Infrastructure. WCML. BS (restricted use).

1206 **C** 31232 **C** 31514 **C** 31526 **C** 31533 **C** 31548 **C**

VPA   Euston – West Midlands Services Class 86.

6101 **I**	86102 **IO**	86103 **I**	86204 **I**	86207 **I**	86208 **I**	86209 **M**
6210 **I**	86213 **I**	86219 **I**	86224 **I**	86225 **I**	86231 **I**	86236 **I**
6240 **I**	86242 **M**	86245 **I**	86248 **I**	86251 **I**	86253 **I**	86256 **I**
6257 **I**	86258 **I**					

VRP   GWML Services Class 43.

3002 **I**	43003 **I**	43004 **I**	43005 **I**	43006 **I**	43007 **I**	43008 **I**
3009 **I**	43010 **I**	43011 **I**	43012 **I**	43015 **I**	43016 **I**	43017 **I**
3018 **I**	43019 **I**	43020 **I**	43021 **I**	43022 **I**	43023 **I**	43024 **I**
3025 **I**	43026 **I**	43027 **I**	43028 **I**	43029 **I**	43030 **I**	43031 **I**
3032 **I**	43033 **I**	43034 **I**	43035 **I**	43036 **I**	43037 **I**	43040 **I**
3041 **I**	43124 **I**	43125 **I**	43126 **I**	43127 **I**	43128 **I**	43129 **I**
3130 **I**	43131 **I**	43132 **I**	43133 **I**	43134 **I**	43135 **I**	43136 **I**
3137 **I**	43138 **I**	43139 **I**	43140 **I**	43143 **I**	43148 **I**	43149 **I**
3150 **I**	43151 **I**	43152 **I**	43163 **I**	43164 **I**	43165 **I**	43166 **I**
3167 **I**	43168 **I**	43169 **I**	43171 **I**	43172 **I**	43173 **I**	43174 **I**
3175 **I**	43176 **I**	43177 **I**	43178 **I**	43179 **I**	43181 **I**	43182 **I**
3183 **I**	43184 **I**	43185 **I**	43186 **I**	43187 **I**	43188 **I**	43189 **I**
3190 **I**	43191 **I**	43192 **I**				

KXS   Stored.

1171 **FO** 31248 **FO** 31413 **O**

RAILFREIGHT DISTRIBUTION SECTOR

MDAT   Tinsley Class 47.

7052 **FD**	47146	47147 **FD**	47206 **FD**	47207 **FD**	47231 **FD** 47238 **FD**
7279 **FD**	47295 **FP**	47296 **FD**	47301 **FR**	47302 **FR**	47305 **FP** 47337 **FO**
7339 **FD**	47347 **FM**	47354 **FD**	47359 **FD**	47367 **FR**	47377 **FD**

MDCT   Tinsley Class 47 (based at Saltley).

7033 **FD**	47051 **FD**	47219 **FD**	47222 **FD**	47236 **FD**	47237 **FD**	47280 **FD**
7285 **FD**	47310 **FD**	47313 **FD**	47316 **FD**	47323 **FD**	47326 **FD**	47362 **FD**

MDDT   Tinsley Class 47 (extended range).

7049 **FD**	47053 **FD**	47085 **F**	47095 **FD**	47114 **FD**	47125 **FE**	47144 **FD**
7150 **FD**	47152 **FD**	47156 **FD**	47186 **FE**	47188 **FD**	47194 **FD**	47200 **FD**
7201 **FD**	47204 **FD**	47205 **FD**	47209 **FD**	47211 **FD**	47213 **FD**	47217 **FE**
7218 **FD**	47226 **FD**	47228 **FD**	47229 **FD**	47234 **FE**	47241 **FD**	47245 **FD**
7258 **FD**	47281 **FD**	47284 **FD**	47286 **F**	47287 **FD**	47290 **FD**	47291 **FD**
7292 **FD**	47293 **FD**	47297 **FD**	47298 **FD**	47299 **FD**	47303 **F**	47304 **FD**
7307 **FE**	47309 **FD**	47312 **FD**	47314 **FD**	47328 **FD**	47330 **FD**	47335 **FD**
7338 **FD**	47344 **FD**	47351 **FD**	47355 **FD**	47360 **FD**	47361 **FD**	47363 **F**
7365 **FD**	47375 **FD**	47378 **FD**				

MDLC Crewe Class 90/0.

90022 **FD** 90023 **FD** 90024 **FD**

MDMC Crewe Class 90/1.

90125 **FD** 90126 **FD** 90127 **FD** 90128 **0** 90129 **0** 90130 **0** 90131
90132 **FE** 90133 **M** 90134 **M** 90135 **M** 90136 **0** 90137 **FD** 90138
90139 **FD** 90140 **FD** 90141 **FD** 90142 **FD** 90143 **FD** 90144 **FD** 90145
90146 **FD** 90147 **FD** 90148 **FD** 90149 **FD** 90150 **FD**

MDNC Crewe Classes 86/6 and 87/1.

86602 **FD** 86603 **FD** 86604 **FD** 86605 **FD** 86606 **FD** 86607 **FD** 86608
86609 **FD** 86610 **FD** 86611 **FD** 86612 **FD** 86613 **FD** 86614 **FD** 86615
86618 **FD** 86620 **FD** 86621 **FD** 86622 **FD** 86623 **FD** 86627 **FD** 86628
86631 **FD** 86632 **FD** 86633 **FD** 86634 **FD** 86635 **FD** 86636 **FD** 86637
86638 **FD** 86639 **FD** 87101 **FD**

MDRL Laira Refurbished Class 37 (based at St. Blazey).

37411 **FD** 37412 **FD** 37413 **FD** 37416 **M** 37669 **FD** 37670 **FD** 37671
37672 **FD** 37673 **FD** 37674 **FD** 37675 **FD**

MDRM Motherwell Refurbished Class 37.

37401 **FD** 37406 **FD** 37409 **M** 37410 **M** 37423 **M** 37424 **M** 37430

MDSR Tinsley Reserve fleet.

37057 **BR** 37239 **FC** 37278 **FC** 37280 **FP** 37373 **FR** 47079 **FD** 47157
47187 **FD** 47196 **FD** 47197 **FP** 47210 **FD** 47276 **FP** 47288 **FD** 47289
47308 **F** 47321 **F** 47352 **C** 47370 **FO** 47371 **FO** 47376

MDTT Tinsley Class 37 (extended range).

37015 **FD** 37019 **FD** 37026 **FD** 37053 **FD** 37068 **FD** 37073 **FD** 37075
37079 **FD** 37107 **FD** 37108 **F** 37110 **M** 37131 **FD** 37154 **FD** 37178
37218 **FD** 37225 **FD** 37238 **FD** 37261 **M** 37298 **FD**

MDWT Tinsley Class 47 (restricted use).

47142 **FR** 47145 **0** 47193 **FP** 47212 **FP** 47221 **FP** 47223 **FD** 47224
47249 **FR** 47256 **FD** 47270   47277 **FP** 47278 **FP** 47283 **FD** 47294
47317 **FD** 47322 **FR** 47345 **FR** 47350 **FO** 47369 **FD** 47379 **FP**

MDYX Stored.

08878   37029 **FD** 37032 **FR** 37070 **FD** 37209 **BR** 37248 **FM** 37252
47019 **FO** 47050 **FD** 47060 **FD** 47063 **FA** 47105   47190 **FP** 47214
47225 **FD** 47306 **FD** 47325 **FO** 47349 **FD** 47358 **FO**

MSCH Rfd Central Hire Pool. Shunters.

08417 **D** 08498   08562   08603   08670   08705   08713
08823   08866

SNA Allerton Shunters.

3397 F	08402 D	08415	08482 D	08485	08489 F	08569
3615	08694	08799	08809	08815	08817 BS	08856
3872 D	08884	08894	08900 D	08902	08913 D	08918 D
3925	08951 D					

SNB Bescot Shunters.

| 3428 | 08535 D | 08543 D | 08610 | 08616 | 08734 | 08746 D |
| 3751 | 08765 D | 08805 FO | 08920 F | | | |

SNC Crewe Shunters.

| 3585 | 08599 | 08633 RX | 08635 | 08692 | 08695 | 08702 |
| 3737 F | 08739 | 08742 | 08784 | 08907 0 | 08921 | |

SNE Derby Etches Park Shunters.

| 3536 | 08690 | 08697 | 08899 | 08956 |

SNL Longsight Shunters.

| 3624 | 08673 IO | 08676 | 08721 0 | 08790 | 08891 | 08915 F |

SNU Carlisle Upperby Shunters.

| 3534 D | 08768 | 08826 | 08827 | 08910 | 08911 D | 08912 |

SNX North West Shunters. Stored.

| 8447 | 08448 | 08601 0 | 08611 | 08613 | 08619 | 08666 |
| 8703 | 08788 | 08844 | 08893 D | 08901 | 08939 | |

SSA Bristol Bath Road Shunters.

| 8410 D | 08483 D | 08643 D | 08668 | 08818 | 08897 D |

SSB Bletchley Shunters.

| 8484 D | 08519 0 | 08567 | 08625 | 08628 | 08629 0 | 08683 |
| 8807 | 08914 | 08927 | | | | |

SSC Cambridge Shunters.

| 8594 | 08685 | 08711 | 08714 | 08757 RX | 08865 |

SSL Laira Shunters.

| 8576 | 08641 D | 08644 I | 08645 D | 08648 D | 08663 D | 08792 |
| 8801 | 08819 D | 08937 D | 08941 | 08953 D | 08954 F | 08955 |

SSM March Shunters.

| 8495 | 08528 D | 08529 | 08538 D | 08540 D | 08580 |

SSN Norwich Shunters.

| 8810 | 08869 G | 08928 FR |

SSO Old Oak Common Shunters.

| 8460 F | 08480 | 08651 D | 08653 | 08825 | 08837 D | 08904 |
| 8944 D | 08947 | 08948 | | | | |

MSSR   Reading Shunters.

08413 **D** 08523   08905   08924 **D** 08946 **D** 09101 **D** 09102 **N**

MSSS   Stratford Shunters.

08393 **D** 08414 **O** 08526   08527 **D** 08530 **D** 08531 **D** 08541 **N**
08542 **F** 08573   08593 **O** 08627   08655 **F** 08689 **O** 08709
08715 **O** 08724   08740 **F** 08750   08752 **C** 08758   08767
08775   08828   08834 **F** 08909   08957   08958

MSSW   Willesden Shunters.

08451   08454   08617   08696 **D** 08842   08873 **M** 08887
08890 **D** 08926   08934

MSSX   South Eastern & Western Shunters. Stored.

08472   08517   08698   08700   08748   08772 **G** 08811

NETWORK SOUTH-EAST SECTOR

NKFE   Infrastructure. South electrification. EH.

37194 **FD** 37220 **FP** 37245 **C** 37375 **C** 37380 **FC**

NKJD   Infrastructure. South. EH & RY Shunters.

03079   08600 **D** 08642 **O** 08649 **G** 08847   08892 **D** 08933 **O**
08940

NKJE   Infrastructure. South. EH.

33008 **G** 33019 **C** 33025 **C** 33030 **C** 33035 **N** 33046 **C** 33051 **C**
33116 **D** 37198 **C** 37274 **C** 37293 **FM** 37377 **C**

NKJH   Infrastructure. South. SU Shunters.

08854   09003   09004   09006   09007   09009 **D** 09010 **D**
09011 **D** 09012 **D** 09016 **D** 09018   09019 **D** 09020   09021
09022   09023   09024 **D** 09025   09026 **D**

NKJL   Infrastructure. South. SL.

73101 **O** 73103 **IO** 73104 **IO** 73105 **C** 73106 **D** 73107 **C** 73108 **C**
73110 **C** 73117 **IO** 73118 **C** 73119 **C** 73129 **N** 73130 **C** 73133 **C**
73136 **N** 73138 **C**

NKJM   Infrastructure. SL. Meldon Quarry duties.

33002 **C** 33057 **C** 33064 **FA** 33109 **D** 33202 **C** 33208 **C** 56031 **C**
56036 **C** 56046 **C** 56047 **C** 56048 **C** 56049 **C**

NKJR   Infrastructure. South. SL (restricted use).

33012   33021 **FA** 33023   33026 **C** 33042 **FA** 33048   33052
33053 **FA** 33063 **FA** 33065 **C** 33207 **FA** 73114 **IO** 73126 **N** 73128 **C**
73131 **C** 73132 **IO** 73134 **IO** 73139 **IO** 73140 **IO** 73141 **IO**

NKJS   Infrastructure. North. SF.

31135 **C** 31181 **C** 31250 **C** 31290 **C** 37013 **F** 37023 **C** 37047 **F**
37055 **FD** 37140 **C** 37244 **FD** 37370 **C** 37371 **C** 37376 **FC** 37379 **C**

KJW Infrastructure. North. SF (weekend work).

1165 **G** 31180 **FR** 31191 **C** 31219 **C** 31268 **C** 37241 **FM** 37242 **FD**

WXA West of England/North Downs Services. LA.

0007 **G** 50033 **N** 50050

WXB West of England/North Downs Services. BM.

3109 **N**

XJX Stored.

3204 **FD** 33206 **FD**

XXA General.

3003 **G**

XXB EMU Transfer. SF.

7526 **BR** 47579 **N** 47702 **N** 47711 **N**

RAIL EXPRESS SYSTEMS SECTOR

XLA Crewe Class 90.

0016 **RX** 90017 **RX** 90018 **RX** 90019 **RX** 90020 **RX** 90021 **FD**

XLB Crewe Class 47 (extended range).

7490 **RX** 47491 **RX** 47500 **RX** 47503 **RX** 47517 **RX** 47537 **RX** 47541 **RX**
7559 **RX** 47562 **RX** 47564 **RX** 47569 **RX** 47573 **RX** 47578 **RX** 47580 **RX**
7581 **RX** 47592 **RX** 47597 **RX** 47603 **RX** 47612 **RX** 47618 **RX** 47630 **RX**
7631 **RX** 47636 **RX** 47641 **RX** 47642 **RX** 47644 **RX** 47653 **RX** 47701 **RX**
7703 **R** 47704 **RX** 47705 **RX** 47706 **PS** 47707 **RX** 47709 **RX** 47710 **N**
7712 **R** 47714 **N** 47715 **N** 47716 **N** 47717 **R** 47774 **RX** 47775 **RX**
7778 **RX** 47824 **M**

XLC Crewe Class 47.

7467 **BR** 47474 **R** 47475 **R** 47476 **R** 47489 **R** 47521 **RX** 47523 **M**
7524 **M** 47528 **R** 47530 **RX** 47532 **RX** 47543 **R** 47557 **RX** 47558 **RX**
7565 **RX** 47566 **M** 47567 **RX** 47568 **RX** 47572 **RX** 47574 **RX** 47575 **R**
7576 **RX** 47582 **R** 47583 **RX** 47584 **RX** 47587 **RX** 47588 **RX** 47594 **RX**
7596 **RX** 47599 **R** 47600 **RX** 47605 **RX** 47615 **RX** 47624 **M**
7625 **RX** 47626 **M** 47627 **RX** 47628 **RX** 47634 **R** 47635 **R** 47640 **R**

XLD Reserve and Stored Locos.

7708 **N**

XLE Crewe Class 86.

6239 **R** 86241 **R** 86243 **RX** 86254 **RX** 86261 **RX** 86401 **RX** 86416 **RX**
6417 **RX** 86419 **RX** 86424 **R** 86425 **R** 86426 **RX** 86430 **RX**

XLH Crewe Class 47 (75 mph maximum).

7463   47471 **IO** 47481 **BR** 47492 **IO** 47501 **R** 47513 **BR** 47519 **BR**
7522 **R** 47535 **R** 47536 **BR** 47547 **N**

REGIONAL RAILWAYS SECTOR

RAJP   Infrastructure. Scotrail. IS Class 37/4.

37427 **RR** 37428 **FP** 37431 **M**

RAJV   Infrastructure. Scotrail. IS Class 37.

37025 **C** 37043 **C** 37069 **C** 37087 **C** 37088 **C** 37099 **C** 37153 **C**
37156 **C** 37165 **C** 37196 **C** 37201 **C** 37211 **C** 37232 **C** 37240 **C**
37255 **C** 37275   37294 **C** 37351 **C**

RBJH   Infrastructure. North East. HT.

37139 **FC**

RBJI   Infrastructure. North East. IM.

37049 **C** 37083 **C** 37144 **FA**

RBJN   Infrastructure. North East. HT (weekend work).

37045 **F** 37059 **FD** 37063 **FD** 37128 **BR** 37202 **FM** 37217   37272 **F**
37285 **F**

RBJW   Infrastructure. North East. IM (weekend work).

31230 **FO** 31247 **FR** 31563 **C** 47346 **C**

RCJC   Infrastructure. North West. CD Class 31.

31188 **C** 31207 **C** 31229 **C** 31233 **C** 31238 **C** 31242 **C** 31255 **C**
31272 **C** 31285 **C** 31306 **C**

RCKC   Infrastructure. North West. CD Class 31/4.

31408   31410 **RR** 31421 **RR** 31432   31439 **RR** 31455 **RR** 31465 **F**

RCLC   Infrastructure. North West. CD Class 37.

37503 **FM** 37504 **FM** 37509 **FM**

RCMC   Infrastructure. North West. CD Class 37/4.

37402 **M** 37407 **M** 37408 **BR** 37414 **RR** 37418 **FP** 37421 **RR** 37422 **F**
37425 **FA** 37429 **RR**

RCRB   Infrastructure. North West. BD.

73001   73002 **BR** 73005 **O** 73006 **BR**

RCWC   Infrastructure. North West. CD Class 31 (weekend work).

31160 **F** 31190 **C** 31263 **C** 31270 **FC** 31282 **FR** 31301 **FR** 31327 **F**
31438

RDDJ   Infrastructure. Central. BS Class 31 (RETB fitted).

31146 **C** 31147 **C** 31158 **C** 31166 **C**

RDJB   Infrastructure. Central. BS Class 31.

31155 **FA** 31174 **C** 31178 **C** 31185 **C** 31237 **C** 31273 **C** 31308 **C**
31468 **C**

OJM Infrastructure: Central. SF (based at March).
1186 **C** 37167 **FC**

OJS Infrastructure. Central. SF.
1187 **C** 37106 **C** 37216 **G**

OJW Infrastructure. Central. BS Class 31 (weekend work).
1128 **FO** 31132 **FO** 31164 **FO** 31317 **FO** 31569 **C**

OKB Infrastructure. Central. BS Class 37.
7114 **C** 37162 **D** 37185 **C**

EJK Infrastructure. South Wales & West. CF.
7141 **C** 37142 **C** 37146 **C** 37158 **C** 37191 **C** 37207 **C** 37263 **C**

EJS Infrastructure. South Wales & West. CF (sandite fitted).
7197 **C** 37230 **C** 37254 **C** 37258 **C**

BR TELECOMS LOCOS

AKB Bescot.
0075 20128 20131 **T** 20187

AKX Stored.
0007 20032 20072 20104 **FR** 20117 20121 20190
0215 **FR**

PRIVATELY-OWNED LOCOS

YPA ARC.
59101 **0** 59102 **0** 59103 **0** 59104 **0**

YPD Hunslet-Barclay.
20901 **0** 20902 **0** 20903 **0** 20904 **0** 20905 **0** 20906 **0**

YPO Foster-Yeoman.
59001 **0** 59002 **0** 59003 **0** 59004 **0** 59005 **0**